S0-BUC-009
12/9/70

MAGNETISM
An Introductory Survey

MAGNETISM
An Introductory Survey

E. W. LEE

DOVER PUBLICATIONS, INC.
NEW YORK

Copyright © 1970 by Dover Publications, Inc.
Copyright © 1963 by E. W. Lee.
All rights reserved under Pan American and International Copyright Conventions.

Published in Canada by General Publishing Company, Ltd., 30 Lesmill Road, Don Mills, Toronto, Ontario.
Published in the United Kingdom by Constable and Company, Ltd., 10 Orange Street, London WC 2

This Dover edition, first published in 1970, is an unabridged republication, with minor revision and a new preface, of the work originally published in 1963 by Penguin Books Ltd. The previous edition did not include the present subtitle.

International Standard Book Number: 0-486-22665-4
Library of Congress Catalog Card Number: 70-13143

Manufactured in the United States of America
Dover Publications, Inc.
180 Varick Street
New York, N.Y. 10014

PREFACE

to the Dover Edition

UPON publication of the new edition of this book I take the opportunity to reiterate its original objective: to explain in simple language to the intelligent non-scientist the basic experiments and thought underlying much of the present understanding of the magnetic properties of matter. The work does not pretend to be a thorough portrayal of the contemporary scene, which changes rapidly with the constant introduction of new materials and new experimental techniques and applications. Indeed, the research field is so active that any author who attempts to delineate it is doomed to be overtaken by events before publication can be achieved.

Given the historical and elementary character of the book, it seemed neither necessary nor wise to attempt an extensive revision of the text. Instead I have incorporated a number of corrections and alterations. The task of correcting the text was made much easier by the letters I received from readers pointing out errors and inaccuracies in the original edition. These letters taught me much and, in several cases, led to acquaintance and friendship. I am most grateful to all those who have written me and have been glad to incorporate most of their suggestions.

CONTENTS

1. Introduction — 7
2. The Early History of Magnetism — 12
3. Later Developments – Electromagnetism — 22
4. A General Synopsis of Magnetic Behaviour — 49
5. The Atomic Theory of Matter — 66
6. Paramagnetism and Diamagnetism — 83
7. Ferrimagnetism — 110
8. Ferromagnetic Domains — 121
9. Magnetic Materials and their Applications — 150
10. The Development of the Permanent Magnet — 167
11. Ferrimagnetism and Antiferromagnetism — 192
12. Magnetism in Scientific Research — 205
13. The Earth's Magnetism — 235
14. Rock Magnetism — 258
    BIBLIOGRAPHY — 270
    ACKNOWLEDGMENTS — 271
    INDEX — 273

CHAPTER 1

Introduction

THE title of this book refers to the magnets of iron that most of us, at one time or another, have found fascinating, and to the property which makes magnets what they are. Our experience of magnets usually extends back to early childhood, although few adults can resist their fascinating behaviour. As children we learn from experience that a magnet is capable of exerting forces on objects made of iron and steel which may be great enough to lift them, and we know that a strong magnet can support a piece of iron many times its own weight. We may have noticed that two similar magnets can repel each other as well as attract. Very often, in our endeavour to find out what a magnet will do, we may be struck by what it will not do. The same magnet which will pick up several dozen steel nails is totally without influence on a brass pin, and we are understandably puzzled. Since our powers of concentration are limited, we do not remain puzzled for long and either return to the steel nails or go off in search of other amusements.

Our acquaintance with magnetism at school is usually first made through the famous experiment in which iron filings are sprinkled over the surface of a magnet laid on a piece of white card, thereby delineating its magnetic lines of force. The impact of this experiment on the pupil's mind is evidently tremendous, for it appears to obliterate the memory of all others. The sum total of magnetism to millions of adults can be adequately summed up in the phrase 'iron filings and all that'. Those who specialize in science at school fare little better, for although they may progress beyond the iron-filings stage, their study of magnetism is usually confined to that of magnetic poles or of bar magnets, and they thus reach a state of knowledge roughly com-

Magnetism

parable with that of Gilbert in 1600. Nevertheless there is an almost universally-held belief that magnetism consists entirely of what 'we did at school'. Consequently the professional physicist or chemist who chooses magnetism as a field for study and research is best advised not to confess the subject of his pursuit if he wishes to avoid the derision that will almost certainly be heaped upon him should he do so. Surely, he will be told, he is not still playing with magnets; everything about magnetism must be known by now. After all, 'we did it at school'. On the contrary, the research physicist maintains that the magnetism taught in schools bears so little resemblance to real magnetism that he and many educationalists would like to see it dropped from the examination syllabuses. He could draw attention to the two major international conferences on magnetism which are held each year and demonstrate the importance and interest in the subject implied by the annual expenditure of large sums of money on magnetic research. He would do better to remain silent, for as likely as not the only response is likely to be a semi-humorous question as to whether in view of this fact there is not the likelihood of a world shortage of iron filings.

It is true that magnetism and magnetic research does not hit the headlines to the same degree as research in nuclear physics. This is hardly surprising, for it is nothing like so glamorous a subject and the advances which have been made in recent times are neither as spectacular nor as immediately understandable as, for example, the discovery of a new nuclear particle or the manufacture of a new element. In spite of this it is undoubtedly true that the pattern of our lives and our material standard of living are affected more by magnetism than they are likely to be by nuclear energy in the next twenty years. If our lives are affected by magnetism why are we so unaware of it? The answer seems to be that applications of magnetism have been coming into use gradually for the last eighty years or so and many of us have grown up with them already in existence. We accept them as part of our everyday lives and do not think to inquire how

Introduction

they work or whether they rely on any special property for their operation.

So far we have been almost exclusively concerned with what magnetism is not, chiefly in order to eradicate any misconceptions which may exist. It is hoped that some idea of what magnetism is about will be found in the ensuing pages. In presenting this account it has been necessary to bear several important factors in mind. In the first place it is not possible to study magnetism and electricity separately, for they are both different manifestations of what is essentially the same thing. The fact that we tend to regard them as separate entities is in some way accidental and stems from the fact that the subjects of electrostatics and magnetostatics arose from the discovery of two minerals, amber and lodestone, whose properties were studied separately.

That these two subjects bear a certain formal resemblance to each other is of little significance when we realize that whereas single electric charges exist, single magnetic poles do not. Indeed this formal resemblance, based itself on a misconception, probably retarded the discovery of the true connexion between electricity and magnetism, which is rightly regarded as one of the greatest achievements of nineteenth-century physics. The concept of magnetic poles is undoubtedly a useful one, and in following the historical development of the subject the author has been forced to devote some time to its exposition. This procedure also conforms with current practice in the teaching of electricity and magnetism in schools and in most universities at the present time. It is possible and indeed desirable to develop the whole subject of electromagnetism without recourse to the idea of magnetic poles. There are indications that this is the way the subject will be taught in future, thereby stressing the unity of the two subjects and avoiding the unfortunate dichotomy which still prevails.

In addition to the conceptual difficulties which may exist in electricity and magnetism, a further difficulty arises from the use of technical terms and mathematics. Every attempt has been made in this book to reduce mathematics to a minimum, often

Magnetism

with a loss in conciseness and exactness of expression, but the same is not true of technical terms, which, however, have been explained wherever necessary. In physics use is made of technical terms and mathematics not to provide a cloak of technical jargon to cover up obscurities, nor to endow scientific work with a kind of esoteric mystery so that it may be comprehended only by other members of the brotherhood. Some of the arguments used in physics are quite subtle and are often very complicated. Consequently the requisite degree of exactness is obtained only by formulating the arguments in terminology which is itself precise and incapable of being misunderstood. Mathematics provides us with the most precise way of formulating statements and mathematical methods are the most precise way of drawing conclusions from them. Consequently a physicist's reasoning often develops along mathematical lines, and he may find some difficulty in putting his reasoning into words. If he should try to do so he must first recollect that words are current coinage and as such liable to be debased or to alter in value. He must therefore use only those words whose meaning is precise and recognized by those with whom he is trying to communicate. These words are technical terms. Every subject has its own list of technical terms, and we must, if we are to understand it, make sure that we appreciate their meaning. Some of these words are used in everyday life without their technical and precise meaning, and we must remember that the everyday meaning may be somewhat different from their technical one. This usage is not common to science, though science uses a large number of technical terms. The word power has a precise meaning in physics which it does not possess in everyday speech. Further examples come to mind quite readily and their implication is clear enough. If we are to understand science we must be prepared to learn the terminology in which its discoveries and laws are formulated.

In what follows, an account is given of some of the more important aspects of magnetism and their application in research and in everyday life. The account is by no means com-

Introduction

plete, and in order to avoid becoming too involved in certain branches of physics, some statements have been invested with an air of generality which they do not rightly possess. Healthy scepticism is essential to the scientific spirit, and if the reader has cause to doubt some of these generalizations and is impelled to consult more advanced texts such as those given in the bibliography, or if in the process of reading he becomes aware that there is still much to be discovered and understood, then this book will have served its purpose.

CHAPTER 2

The Early History of Magnetism

THERE is considerable justification for the belief that modern science stems from Greece and has its origin in the speculation of her philosophers. These individuals, men of thought rather than action, based their speculation on observation and their experience of the world as revealed by the five senses. The properties of matter did not escape their attention, and they recognized three different forms of matter exemplified by earth, water, and air, namely solids, liquids, and gases. It was appreciated too that some substances have certain properties which distinguish them from others in the same class: wood and iron are both solids, yet wood is combustible and iron is not; gold is yellow and copper is red, and so on. But amongst the solid substances known to the ancients there were two whose properties singled them out as being in a class almost of their own. These were two minerals, amber and lodestone. Amber has the power of attracting light bodies, such as small pieces of tissue paper, when it has been rubbed, and lodestone attracts iron without previous mechanical treatment. When these two substances were discovered, and by whom, we shall never know, but they were certainly well known to the Greeks and their behaviour was the subject of some speculation. We should hardly expect otherwise, for the forces exerted by amber and lodestone act over considerable distances and in this respect are quite different in behaviour from the purely mechanical forces associated with pushing and pulling in which there is intimate contact between the object providing the force and that which is moved. But by and large these two substances must have remained mere curiosities for about two thousand years. Then at some time in the early middle ages – the exact date is unknown – someone discovered that a piece of lodestone, if held in such a way as to allow it to rotate freely, probably by attaching it to a light object such as cork or straw, floating in

The Early History of Magnetism

water, always set in one particular geographic direction, and this fact was used in navigation for indicating direction at sea. The inventor of the compass is not known. It was at one time believed that it originated in China and was brought by the Arabs to the Mediterranean where it was first seen by the Crusaders, but recent research has cast doubt upon the truth of this. The Chinese certainly were aware of the directive property of lodestone, but its application as a means of indicating north and south seems to have been confined to use on land. The first report of the use of the compass for navigation comes from a Chinese writer Chu Yu who noticed that 'foreign' sailors were employing it on the ships sailing between Canton and Sumatra. The earliest European reference to the compass is in a work by Alexander Neckam (1157–1217), an English monk of St Albans, but he does not claim it to be a new invention. This alone implies a certain familiarity with the compass not only on his own part but on that of his contemporaries as well, and it seems likely that it was quite well known in parts of Europe and Asia by the beginning of the thirteenth century.

In 1269 Petrus Peregrinus de Maricourt, a French crusader, not only gave the first detailed description of the floating compass but described in some detail a new pivoted compass of the type with which most of us are reasonably familiar. This was not all, for he went on to make a discovery of great importance. Taking a piece of lodestone which had been fashioned into a spherical shape he took a pivoted needle and explored the surface of the sphere. At any point on the sphere he found that the needle would set itself in one direction only, and he marked on the sphere a line along the direction which the needle set itself. He did this repeatedly on other different parts of the sphere. When the entire surface of the sphere had been investigated in this way it was covered with lines, and it was immediately apparent that the lines girdled the sphere in precisely the same manner as the lines of longitude girdle the earth. In particular all these lines passed through two points at opposite ends of the sphere just as the lines of longitude pass through the north and south poles of the earth. By analogy he called these two points the poles of his spherical magnet. As a result of further experiments

Magnetism

he observed that the way in which magnets set themselves when free to move, and the way they attracted objects made of iron, depended solely on the position of these poles. Evidently these poles were the seat of the magnetic power. At this stage it is necessary to draw attention to one important feature, although its significance could not be appreciated at the time. Every magnet was observed to possess *two* poles, never one, with the magnetic lines forming a continuous path through them.

After this time little progress was made until the work of William Gilbert (1540–1603), a native of Colchester. Gilbert extended the observations of Peregrinus and amongst other things explained the fact that magnets always take up one particular orientation at any point on the earth's surface. He realized that the earth is itself one huge magnet with its magnetic poles situated at or near the geographical north and south poles. Consequently every magnet on the earth's surface behaves just like the needle on Peregrinus's lodestone globe. But he went further, because once it was realized that the earth has two magnetic poles all observations could be explained by the assumption that the forces acting on a magnet are concentrated solely at the poles, and that two north-seeking poles repel each other, while two unlike poles, i.e. a north-seeking pole and a south-seeking pole, attract. In this way the process of clarification and abstraction proceeded. All the power of a magnet is concentrated at its poles, and the forces between magnets may be seen simply as the resultant of the forces between each of the poles. But we must not lose sight of the fact that every magnet has two poles, for it is this which distinguishes magnetism from electricity – to whose early history we must now turn.

The fact that amber, when rubbed by silk, flannel, fur, or similar materials, acquires the power to attract light objects has already been mentioned. Gilbert showed that this power was not exclusive to amber, as had been thought to be the case for centuries, but was possessed by a number of substances, among which he mentioned glass, sulphur, and sealing wax. It will be realized that these substances belong to a class which an electrician of today would describe as insulators. The force which

The Early History of Magnetism

these substances exerted as a result of friction was termed electric by Gilbert, and although he realized that there were certain similarities between electric and magnetic forces he was not slow to observe that there were equally important distinctions between them. The force exerted by lodestone on iron substances is inherent in the nature of lodestone and needs no friction to bring about its appearance. On the other hand lodestone acts on iron only, whereas electrified bodies attract everything.

Progress in understanding the nature of electric forces or electricity was slow. In 1729 Stephen Gray noticed for the first time that the power possessed by an electrified body could be passed from one such body to another un-electrified body by making connexion between them with certain substances, amongst which the metals were by far the most efficient. These substances were given the name non-electrics or, in modern terminology, conductors. The next important discovery was made by du Fay about 1734. It had been noticed that in a stranded bifurcated electrified body such as a piece of gold leaf hung over an insulated support, a force of repulsion exists between one part of the leaf and another. Precisely the same phenomenon may sometimes be observed when combing our hair, for the friction between comb and hair electrifies the hair, as a result of which it often refuses to lie down until the state of electrification of the hair is removed, for example by laying a hand gently on it. Thus similar objects, similarly treated, repel each other. Our hair 'stands on end' after being combed because every individual hair, having been rubbed by the same comb, becomes electrified in the same manner and tries to repel every neighbouring hair. Du Fay noticed for the first time that when two dissimilar substances, glass and copal (a resin used in making varnish) were both rubbed in the same way with a piece of flannel they became electrified in such a way that there existed a force of attraction between them. When a body becomes electrified we say that it acquires electricity, and in order to account for both attraction and repulsion between them there must be two kinds of electricity just as there are two kinds of magnetic poles.

It was left to the great American scientist Benjamin Franklin

Magnetism

to show in 1747 what really occurred in the process of electrification by friction. Franklin showed that when glass is rubbed with silk, electricity is not created on the glass by the act of rubbing, but that the friction merely causes a transference of electricity from glass to silk. The silk gains as much as the glass loses, and so the total quantity of electricity in the system remains constant. Instead of referring to electricity lost and electricity gained it is convenient to introduce the terms positive and negative electricity. By this means the quantitative development of the subject is made easier. If in the process of rubbing glass with silk, the glass acquires an amount of charge denoted by $+Q$, then the charge acquired by the silk (note the change of wording here; having introduced the idea of negative electricity it is no longer necessary to regard it as something left when the positive electricity is removed) must be $-Q$, so that the total charge on glass and silk together is $+Q + (-Q) = 0$. Thus what Franklin had shown was that neutral (i.e. unelectrified) bodies are those which contain equal amounts of positive and negative electricity. Friction between bodies merely causes an excess of (say) positive electricity on one body by removing from it some of the negative electricity and depositing it on the other. Such bodies are then said to be charged with electricity or to carry an electric charge, and the forces of attraction and repulsion between electrified bodies are the forces between the electric charges that they carry. Just as is the case with magnetic poles, the force between two charged bodies acts even though the bodies are not in contact. However it is a matter of common experience that an electrically charged body will pick up small pieces of paper only when it is held very near them. Likewise the attraction of a magnet for iron is great enough to be felt only when magnet and iron are placed close together. If we are to interpret the forces between charged bodies as the force between the charges, these observations imply that the force between charges must depend in some way on the distance between them. The problem was taken up by Joseph Priestley, better known as the discoverer of oxygen, who showed by an indirect experiment that the force between charges varies inversely as the square of the distance between

The Early History of Magnetism

them, i.e. the force varies in the same manner as the force of gravitation between massive bodies. It was left to the great French scientist Charles Coulomb (1736–1806) to confirm this law by a direct experiment, i.e. by straightforward measurement of the force between two small pith balls, charged with electricity, and its dependence upon their distance of separation.

Meanwhile precisely the same law of force between magnetic poles had been discovered by John Mitchell (1724–93) at Cambridge in 1750, and was later confirmed by Coulomb both for the attraction between unlike poles and the repulsion between like poles.

The inverse square law of force between electrical charges and magnetic poles implies that when the distance between two charges is doubled the force is decreased by 2^2, i.e. by 4, when the distance is trebled the force is reduced to one-ninth, and so on. This explains why magnets only seem to attract or repel each other when they are close together. When only one inch apart the force between two magnets may quite well be large enough to be felt by hand. If their distance apart is increased to one foot this force is so diminished as to escape observation.* The inverse square law occurs in other branches of physics besides electricity and magnetism, notably in gravitation, for it is found that two bodies attract one another with a force proportional to their masses and inversely as the square of their separation. Unless the two gravitating bodies are very large, for example the earth and the sun, gravitational forces are small and in laboratory experiments negligible. Gravitational forces, apart from being usually smaller than electrical and magnetic forces, differ from them in one important respect, for whereas electrical and magnetic forces can be attractive or repulsive, gravitational repulsion is unknown.

With the concept of magnetic poles and electric charges of either sign and the law of force between them firmly established,

---

* The inverse square law is for individual magnetic poles or electrical charges. Magnets, having two poles, one at each end, exert a force on each other which obeys an inverse fourth power law so that if their separation is increased from one inch to one foot, the force between them is reduced to $(\frac{1}{12})^4$ i.e. to about $\frac{1}{20,000}$ of its original value.

Magnetism

the time was ripe for the mathematical development of the related subjects of magnetism and electricity. Strictly speaking the state of knowledge of both subjects was confined to the properties of magnets and electric charges at rest, or to those subjects which we now term magnetostatics and electrostatics respectively. In a very short time the foundations of the mathematical theory of both subjects had been firmly laid by Simeon Denis Poisson (1781–1840), known also for his contribution to the theory of elasticity. Poisson's mathematical theory, which is remarkable for its completeness, was, in spirit, much in advance of the experimental side of the subject, and it still stands today, forming the basis of the electrostatics and magnetostatics which are studied today not only for their intrinsic interest but because they form one of the starting-points from which the whole subject of electricity and magnetism may be logically developed.

From the middle of the eighteenth century onwards the situation becomes much more complicated. Until that time electricity meant no more than electrostatics. Thereafter it began to mean electric currents as well. As mentioned previously, electric charge may be conveyed from a charged body to an uncharged one by connecting them with a metal wire. If this is done, electric charge flows from the charged body along the wire until a kind of electrical equilibrium is established in a manner very similar to the passage of water from a filled vessel to an empty one if the two are connected by a tube. This flow of electric charge is called an electric current. Of course in the example just quoted electrical equilibrium is attained very rapidly and then the current ceases to flow. Once equilibrium has been established the two bodies are said to be at the same electrical potential. This concept of electrical potential is very similar to the notion of a head of water. Suppose we have two cans of different sizes placed on a level table. These represent our two bodies. If they are of different sizes the same amount of water will rise to different levels in each. The amount of water contained by each is analogous to the amount of electric charge on each body. The height of the water determines the pressure at the bottom of the vessel; alternatively it is a measure of the

The Early History of Magnetism

amount of energy which could be extracted from the water contained in this way, for example by allowing it to flow out of a tube near the bottom and making it drive a water turbine as is done in hydro-electric power stations. Notice that the smaller the capacity of the vessel the higher is the level for a given quantity of liquid. In electrostatics a body of small electrical capacity is one which is raised to a high electrical potential by an electric charge, and just as water flows from the higher level to the lower when the two vessels are connected, so electric charges flow from a high potential to a low potential. When the water levels are equal no more water flows, and when the potentials are equal no more charge flows.

If we require to maintain a continuous flow of water we must pump water out of one vessel and place it in another, thereby maintaining a continuous difference in the water levels in the two vessels. The pump will consume power, since energy has to be supplied to the pump in order that it may do the work of raising water to a higher level, but if we arrange for this to be done we shall have a continuous flow of water in the pipe connecting the two vessels. Similarly in the electrical case we must have a kind of electricity pump for raising the electrical charge from a low to high potential, or in other words means of maintaining a potential difference between the two bodies in order to maintain a continuous flow of electricity along the wire. 'Electricity pumps' exist; we nowadays prefer to speak of them as cells or batteries, and one such cell, consisting of two plates of dissimilar metals immersed in a fluid (a dilute acid) had in 1786 been accidentally discovered by Luigi Galvani (1737–98).

Galvani himself mistakenly attributed the action of the cell to what he himself termed animal electricity (the discovery had been made during an investigation involving the contraction of the muscles in the leg of a frog, and Galvani was here guilty of mistaking cause for effect). The true nature of the device was established by Alessandro Volta (1754–1827), who insisted that the electricity is caused by the action of moisture on the two different metals. We are not going to be concerned with the production and maintenance of electric current, but rather with the effect it produces, so that it is not necessary to

Magnetism

discuss Galvani's discovery any further. The important thing is that means were now available for keeping electrical charges in continuous motion and so the effect of moving charges could now be readily investigated.* Just how important these effects were, we shall now see.

Up to this time, the only connexion between electrical charges and magnetic poles was the fact that they each exert forces on each other which obey the inverse square law, and except for this formal analogy between them the subjects of electricity and magnetism were treated separately and independently. Whatever connexion might exist between them could only have arisen from intelligent guesswork and speculation. It is true that there were pointers, notably a growing belief that a connexion might exist between lightning (which Franklin had shown to be electrical in nature) and magnetism. Belief in this view was strengthened by a number of observations of iron objects becoming magnetized during thunderstorms. However, it was not until 1820 that the connexion was eventually established by Hans Christian Oersted (1777–1851) after a long series of abortive experiments by himself and others. Oersted discovered that a wire carrying an electric current was able to deflect a pivoted compass needle when the wire was placed parallel to the original direction of the needle. Since such a deflection could only be brought about by magnetic force, it follows that the wire carrying an electric current is associated with a magnetic effect. The connexion between electricity and magnetism had at last been found; the two are linked by the simplest of physical phenomena, that of motion—motion of the electric charges along the wire. It seems strange that such a simple connexion should have taken such a long time to find,† but we must remember that the physicists of the time were groping almost blindly with very little previous knowledge of any use to guide them along

---

* Actually it was not until 1833 that Michael Faraday succeeded in proving that the static charges produced on electrified bodies by friction are identical with the charges moving in a wire which constitute an electric current.

† Oersted announced his intention of investigating the effect of electricity on the magnetic needle in 1807.

The Early History of Magnetism

suitable avenues of inquiry and with only the most primitive pieces of apparatus with which to carry out their experiments.

Oersted's discovery was of the utmost importance, perhaps the most important single discovery in magnetism since Peregrinus's discovery of magnetic poles. For it marked the end of an era, an era in which the study of magnetism meant only the study of magnetostatics, with nothing but a formal connexion with electricity. With the new discovery all this was changed and as magnetostatics rapidly receded into the background a new branch of physics began to take its place as an object of serious study. This subject, involving the mutual interaction between magnet and electric current, is now usually known by the name electromagnetism. Henceforth magnetism and electromagnetism are virtually synonymous.

Electromagnetism advanced very rapidly in the subsequent twenty years, and did so for two very good reasons. As a subject it attracted the attention of some of the most illustrious scientists of the time. This alone would have been sufficient. But it is possible to see an additional reason for its rapid growth. Oersted's discovery was made at a time of considerable technological upheaval. Communication between towns was becoming more widespread with the building of new roads, bridges, and canals, and, with the advance of the steam locomotive, more rapid. It was an age in which men looked towards the future with hopeful apprehension. For the first time they began to see that the scientific discoveries which were being made might one day be put to some practical use for the benefit of mankind. This was a new viewpoint, and it can hardly have failed to impress and inspire the scientists of the time, for most of these were not only men of great intellectual ability but men of vision and aspiration who saw the possibility of practical outcomes of their discoveries with the hope of improving conditions of life for their fellow men. This period of growing social awareness, the same as that in which romanticism was becoming a force in music and poetry, signals the first awakening in the minds of the scientists of the time that their discoveries might ultimately be of use (as distinct from being of interest) to others.

CHAPTER 3

Later Developments – Electromagnetism

OERSTED himself never followed up his original discovery, which was described at a meeting of the French Academy some two months after its original announcement. The subsequent development took place largely at the hands of the French school of physicists, notably of Jean-Baptiste Biot (1774–1862), known also for his work on optics, Félix Savart (1791–1841), who also made important contributions to the study of sound, and André-Marie Ampère (1775–1836), who more than any other single individual was responsible for laying the foundations of electromagnetism. Before describing these developments it is necessary to pause for a while and explain some of the concepts and terminology used in magnetism.

The seat of the magnetic attraction between magnets is supposed to lie in the poles. By convention the poles of a magnet are called north and south, the north pole being that which points north. This is in some ways unfortunate, since we know that like poles repel each other, and this implies that the geographic north pole is actually a magnetic south pole. However, in spite of this anomaly, the terminology persists. Sometimes a north pole is referred to as a positive pole and a south pole as negative, and this is in many ways less confusing.

Every magnet possesses two unlike poles, and every attempt to produce single magnetic poles of one kind has been unsuccessful. In spite of this we shall often find it convenient to speak as though isolated single magnetic poles did exist and to describe their effects. Since the force between magnetic poles decreases rapidly as the distance between them is increased it is not too difficult to investigate in practice the forces between two poles only, since by using very long magnets one can arrange to bring one pole of one magnet close to one pole of another and examine their effects, while the two remaining poles can be kept at a distance great enough to ensure that the forces they exert

Later Developments – Electromagnetism

are negligible. In this way the fact that the force between two poles obeys the inverse square law was established. But the force also depends on the strength of the poles, just as the force of gravity between two massive bodies is proportional to the product of their masses. Suppose we have two poles A and B, and measure the force between them when they are a certain distance apart. If we now replace B with another pole C at the same point and find the force between A and C to be twice the force between A and B, we say that the pole strength of C is twice that of B. We could repeat this procedure using different poles D, E, F, and so on, and measure the strength of each pole in terms of the strength of A. Comparison of the force between any two poles thus measured, say B and E with C and D, would show that the magnitude of the force, in addition to being inversely proportional to the square of their distances apart, is also proportional to the strength of each of the poles. Mathematically this means that we can write the law of force between two poles of strength m_1 and m_2 as

$$F \text{ is proportional to } \frac{m_1 \times m_2}{d^2}$$

where d is the distance between them. So far we have elected to measure the strengths of all poles in terms of one, arbitrarily chosen, but this is not very satisfactory, since there is no way of being certain of the constancy of the strength of the pole thus chosen; and so we adopt a new procedure. We choose as a unit pole that which when placed at unit distance from a similar pole experiences unit force. On the metric system of units used by scientists the unit of length is the centimetre, that of mass the gram, and that of time the second (this is usually referred to as the C.G.S. system). Thus on this system unit length means one centimetre. The unit of force is called the dyne; this is a small force roughly equal to the force of gravity acting on $\frac{1}{1000}$ of a gram. Our definition of pole strength now reads 'a unit magnetic pole is that which when placed 1 cm. from a similar pole experiences a force of 1 dyne' and the expression for the law of force between poles now becomes

$$F = \frac{m_1 \times m_2}{\mu d^2}$$

23

Magnetism

μ being a constant called the magnetic permeability of the intervening medium. On our system of units $\mu = 1$ only for a vacuum. But for air μ is only about 1·0000004 (the fact that it is as large as this is due entirely to the presence of oxygen without which μ would be about 1·000000001), so that for most practical purposes no serious error is likely to arise from putting $\mu = 1$. The unit of pole strength defined above is quite a large unit. For example the pole strength of a small compass needle is likely to be only about one-tenth of the unit pole defined in this way.

It is sometimes inconvenient when dealing with the forces between magnets to refer everything back to the forces between poles. Indeed the very fact that an electric current can exert a force on a magnet is an indication that some new concept is necessary. If we had two magnetic poles we should observe a force between them – suppose this is an attractive force. Now if one of the poles were placed in a sealed box we should observe that the other pole experiences a force whose strength depends upon where the pole is placed. Without knowing that the force was due to another pole inside the box we could describe the situation by saying that something acts and exists at all points where the pole experiences a force. This something we call a magnetic field. We should notice that this is simply a term used to describe a certain situation, namely that of a magnetic pole experiencing a force. Nevertheless physicists believe in the physical reality of a magnetic field even if there is no magnetic pole present to provide proof. The magnetic field at any point can be defined quite simply as the force which would be experienced by a unit north pole placed at that point. The magnitude of the force gives the magnitude of the field, and the direction in which the pole would move if it were free to do so gives the direction of the field at that point. The unit of magnetic field strength is called the oersted. The earth's own magnetic field is about 0·6 oersted over most of England, while the field between the pole tips of a really strong permanent magnet is usually about 5,000 oersteds. Magnetic fields of over a million oersteds have been produced, but only for short periods of time.

Later Developments – Electromagnetism

A magnetic field which has the same strength and direction at all points is said to be uniform. Actually no perfectly uniform field can exist, but it is often possible to have a magnetic field which is substantially uniform over quite a large area. The earth's field for example is for all practical purposes uniform over any reasonable area.* A magnet placed in a uniform field

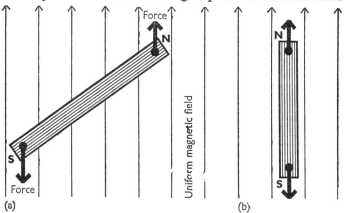

Figure 1. A magnet placed in a uniform magnetic field experiences a twisting effect because of the equal and opposite force on its poles. (a) The twisting force is zero only when the magnet lies along the magnetic field (b) and this is the position in which it will come to rest if freely suspended. Note that a uniform field is represented by a set of equally spaced parallel lines.

experiences no force of translation (i.e. one tending to move it parallel to itself). This is easy to understand because the force exerted by the field on the north pole is exactly equal and opposite in direction to that on the south pole. The forces do not, in general, act in the same straight line. Equal and opposite forces acting in this way are said to constitute a couple, which tends to rotate the magnet until it is in line with the field, when the couple† disappears. This turning effect on a magnet is

---
* The area may be very small in buildings in which a large amount of steel has been used in the construction.

† Strictly, the moment of the couple.

Magnetism

illustrated in Figure 1. If the magnet is freely suspended or pivoted it will come to rest when it lies along the direction of the magnet field so that there is no turning effect acting upon it. This is very useful, since it means that we may use a small pivoted magnet, usually called a plotting compass, to find the direction of the magnetic field at any point. If the compass is moved in such a way that the south pole of its needle is at the point originally occupied by its north pole and the procedure

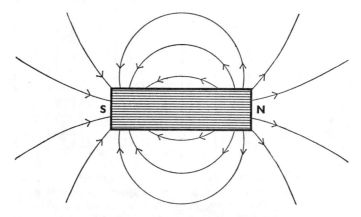

Figure 2. The lines of force due to an isolated magnet.

is repeated many times we eventually trace out a curved line which gives the direction of the field at any point. Such lines are termed lines of force. The lines of force associated with a bar magnet are shown in Figure 2. A simpler though rather more messy way of obtaining the lines of force is to place the magnet on a piece of white paper and sprinkle iron filings round the magnet. Under the action of the magnet each filing becomes a small magnet which behaves like a small compass needle. These can usually be made to turn by gently tapping the table on which the magnet is placed. This, probably the most familiar experiment with magnets, is not wholly to be recommended, since it is usually difficult to remove the iron filings which get stuck on the magnet.

Later Developments – Electromagnetism

When a magnet is placed in a uniform magnetic field we have seen that it is subjected to a turning force. The maximum turning force is when the magnet is at right angles to the magnetic field. If the pole strength of the magnet is m and the field strength is denoted by H, the force on each pole is $m \times H$. The turning effect, usually called a turning moment, about the centre is equal to twice this force (since there are two poles) multiplied by the distance l from the pole to the centre. Mathematically:

$$\text{Turning moment} = 2 \times m \times H \times l$$
$$= 2ml \times H$$

For a given magnetic field strength the turning moment depends only on the product of the pole strength of the magnet multiplied by its length (or more strictly by the distance between the poles, since the poles of a magnet are usually situated somewhere inside it). This quantity is called the magnetic moment of the magnet. Its importance lies in the fact that at distances from the magnet which are large in comparison with its length the field at any point due to the magnet is determined only by its magnetic moment and the distance from the centre of the magnet to the point. Now suppose we decrease the distance between the poles, at the same time increasing the pole strength so that the magnetic moment remains constant. If we carry out this process to the final limit we have a system of two very strong poles (strictly infinitely strong) separated by a very small (strictly infinitely small) distance. Such a system is termed a magnetic dipole, and its magnetic field at any point is governed solely by the magnetic dipole moment and the distance between it and the point in question. The magnetic lines of force due to a magnetic dipole are very similar to those due to a magnet of finite length.

We are now in a position to appreciate more fully the significance of Oersted's discovery. With no current in the wire the magnetic needle points north in order to lie along the lines of force due to the earth's field. When the current is switched on the magnet rotates. But a freely pivoted magnet always comes to rest along the magnetic lines of force. Evidently these have

Magnetism

now been deflected from their original south and north direction. How can this happen? Everything becomes clear when we remember that a magnetic field has not only magnitude but direction, and when two magnetic fields are superimposed the resultant field differs not only in magnitude but in direction from the two original fields. Quantities like magnetic field which require direction as well as magnitude for their complete description are called vectors. Other examples of vectors are force (push upwards), velocity (20 knots in a south-east direction), acceleration, and so on. Vector quantities may be added together, but only in such a way as to take account of both their magnitude and their direction.

It should now be clear that Oersted had really discovered that an electric current flowing in a wire produces a magnetic field. His compass needle set itself along the resultant sum of the earth's field and the field due to the wire. From the fact that the compass needle was deflected in opposite directions when the wire was placed above and below the needle it was established that the magnetic lines of force due to the current formed concentric circles with the wire as centre (Figure 3a). The direction of these

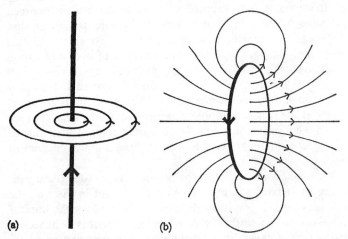

Figure 3. The magnetic lines of force due to an electric current (a) in a straight wire and (b) in a closed circular loop.

Later Developments – Electromagnetism

lines of force can be readily established with the aid of a plotting compass or by the use of iron filings. That the magnetic field due to a current is essentially the same as that of a magnet was demonstrated by Arago, who showed that it is capable of inducing magnetization in iron. Arago had in fact made the first electromagnet.

So far attention had been concentrated on the magnetic effects of current in straight wires. Ampère and also Biot and Savart next directed their attention to currents flowing in closed loops of wire. They reached conclusions identical in nature, although somewhat different in formulation, but it was Ampère who realized the time significance of the result for the theory of magnetism. If we look at the lines of force produced by a current in a circular coil of wire (Figure 3b) we see at once the very strong resemblance between them and the lines of force due to a short magnet. Ampère went a step further. He showed that the field at any point due to a circular current depends only on the current and the distance between the centre

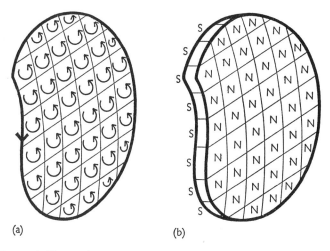

Figure 4. Illustration of the equivalence of a closed loop of wire carrying an electric current (a) and a magnetic shell (b). One must imagine the mesh dividing the area of the loop to be very much finer than that drawn.

Magnetism

of the circle and the point in question, provided this distance is large compared with the diameter of the circle. Moreover the field varies with distance in the same manner as for a magnetic dipole, namely as the inverse cube. Ampère rapidly realized that a current flowing in a tiny circle was exactly equivalent to a magnetic dipole as far as its external magnetic field is concerned. But the idea of a magnetic dipole is something of an abstraction, and the word tiny in the previous sentence really means infinitely small. For currents flowing in conductors of finite size he proceeded in the following way. Imagine a closed loop of wire carrying an electric current to be divided up by a fine network into a large number of small circuits as shown in Figure 4. If the current in each mesh is the same in magnitude as that of the original closed circuit, the current cancels everywhere except on the boundary which is the original circuit. According to Ampère the current in each mesh is equivalent to a magnetic dipole provided the mesh is made sufficiently fine. So our original closed circuit is exactly equivalent to a thin layer of magnetic dipoles arranged parallel to each other and with their axes everywhere at right angles to the layer. Such a layer with one face all north poles and the other all south poles is termed a magnetic shell.* The field due to a magnetic shell may be calculated. It turns out to depend only on the strength of the shell and the shape of its rim. By strength of the shell we mean the magnetic dipole moment per square centimetre of surface. According to Ampère this shell is exactly equivalent to a current-carrying coil whose perimeter coincides exactly with that of the shell. One consequence of this which we shall mention only briefly is that this equivalence must mean that there is a proportionality between the current flowing in a closed circuit and the strength of the equivalent magnetic shell, and it is possible to define a unit of electric current in such a way as to make these two quantities equal. If this is done then we have one very important result (for the current is then equal to the

---

* With the advent of new permanent magnet materials (see Chapter 10) it is now possible to produce discs magnetized at right angles to their plane. Such discs approximate very closely to the concept of an ideal magnetic shell, which however is strictly infinitely thin.

Later Developments – Electromagnetism

shell strength, which is the dipole moment ÷ area), that a current flowing in a small circle is equivalent to a magnetic dipole of strength equal to current × area of circle. We shall need this result later when we come to deal with the magnetic properties of atoms and molecules.

The concept of the equivalent magnetic shell is undoubtedly a useful one, but Ampère perceived that its implications might be even more important. For if a magnetic shell is exactly equivalent to a system of minute circulating currents, might not also the magnetism of a permanent magnet be due to the same cause? If this were so it would explain why a magnet always has two poles and that whenever a magnet is cut in half opposite poles appear at the faces of the cut. On this hypothesis a magnet consists of a large number of currents all circulating in the same sense. Each of these is equivalent to a magnetic dipole, so however we subdivide our magnet we can never have less than one circulating current, with the magnetic lines of force emerging in such a way as to make one face of the circle a north pole and the other a south pole. Modern atomic theory would, in fact, endorse this viewpoint in principle, as we shall see in a later chapter.

What we can also now see is something less fortunate, namely that our description of magnetism and magnetic phenomena so far given is based upon a misconception – the idea of magnetic poles. There is in fact no need to base our theory on the idea of both dipoles and circulating currents, since these two are equivalent, and since modern atomic theory has confirmed in essence the presence of circulating currents in matter, it means that we should regard the concept of magnetic poles as no more than a convenient fiction, a concept useful in explaining certain phenomena but one we should be careful not to believe in too literally. No doubt it would be better, and in the final analysis less confusing, to dispense with magnetic poles altogether, and this can certainly be done, for it will be remembered that basically the idea of poles was introduced in order to explain the observed tendency of a magnet to twist in a magnetic field, in terms of forces, namely those on the poles. Had we been content to work throughout in terms of couples instead of forces

Magnetism

the need to introduce the concept of poles need never have arisen. However this was not the way the subject developed. Science advances by the process of postulation and confirmation or denial by experimental test. Many hypotheses put forward in the eighteenth century have had to be abandoned altogether, notably the phlogiston theory of combustion. It is the hypotheses which are only half wrong that live longer, and as a result the ideas of single magnetic poles (which cannot really be defended upon any grounds, except those of expediency) and dipoles linger on today, although neither is very satisfactory and neither is necessary. Part of this arises from the fact that electricity and magnetism – for the time has now come to put them in the correct order – are young sciences. By the middle of the last century, when these ideas about the nature of magnetism were beginning to gain acceptance, other branches of physics, mechanics in particular, but light and heat as well, were in a state of almost final development. Definite proof of the existence of circulating currents as the basic origin of magnetic behaviour did not come until the beginning of the twentieth century, so perhaps we have not yet had time to see magnetism in its proper perspective. Until that time comes it will still be useful to talk of poles and dipoles.

Shortly after Oersted's discovery the question was asked whether, in view of the fact that the magnetic field produced by a current exerts a force on a magnet, the same current experiences a force by virtue of its being in a magnetic field, namely that of the magnet. Such force might be expected from Newton's third law of motion, usually stated in the form that 'action and reaction are equal and opposite'. This effect was sought and indeed found by Oersted himself. Experiment shows that the force on a straight wire placed in a uniform field is equal to the product of the current flowing in the wire, the length of the wire, the magnetic field strength, and the sine of the angle between the latter and the direction of the current flow. Consequently the force is zero if the current flows along a line parallel to the magnetic lines of force and is a maximum when it is perpendicular to them. It turns out that the force acts

Later Developments – Electromagnetism

in a direction at right angles to both.* If we are careful to use a magnetic field with the correct configuration it is possible to use this fact to obtain continuous motion of a conductor. This was first achieved by Michael Faraday (1791–1867), probably the most gifted experimenter physical science has ever known.

One type of motor developed by Faraday is shown in Figure 5.

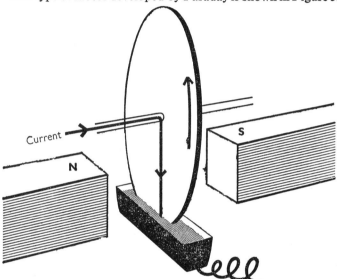

Figure 5. Faraday's second motor. The current flows between the axle and the pool of mercury into which the wheel dips, and the force is always tangential to the rim of the wheel, which is thereby set into rotation.

Here the conductor is a solid wheel, electrical contact being obtained by allowing its rim to dip into a pool of mercury. The magnetic field is uniform (or nearly so) and applied between the axle of the disc and the mercury contact. If a current is passed

* The direction of the force is given by a simple rule due to Fleming. If the thumb and first two fingers of the left hand are held at right angles to each other and the first finger represents the direction of the field, the second finger shows the current and the thumb points along the direction of the force.

Magnetism

from the axle to the perimeter of the disc through the mercury contact this current is always perpendicular to the magnetic lines of force, and so a mechanical force is exerted sideways which tends to rotate the disc. The disposition of magnetic field and current is not upset by the disc's rotation and so the motion is continuous.

If the current is made to flow in a closed loop of wire in a uniform magnetic field the situation is somewhat different. For we know that such a current is equivalent to a magnetic shell,

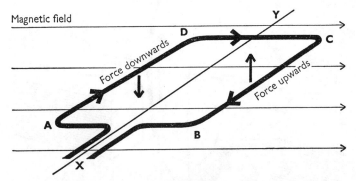

Figure 6. A rectangular coil of wire carrying a current experiences a turning force if placed in a magnetic field.

and a magnetic shell, like the dipoles of which it is made up, experiences no translational force in a uniform magnetic field but only a couple tending to twist it. We can see how this twisting effect arises by considering a small rectangular coil $ABCD$ in Figure 6. The currents in AB and CD are parallel to the magnetic lines of force and consequently there is no mechanical force on them. The current in BC is perpendicular to the magnetic field and it experiences a force perpendicular to both, in this case upwards. So too does the current in AD, but as far as the magnetic field is concerned the direction of the current in AD is in the direction opposite to that in BC and so the force is in the opposite direction, i.e. downwards. Consequently the coil tends to rotate about the axis XY. If it is completely free to rotate it will do so until the plane of the coil is perpendicular to

Later Developments – Electromagnetism

the direction of the magnetic field. The twisting force is now zero, since the forces on BC and DA are in the same straight line. In any real coil, however, its inertia will carry it a little way further round, and if at this stage the current in the coil is reversed it will rotate a further half revolution. If the current is now reversed again the coil rotates a further revolution and by arranging for the current to be always reversed when the coil is at right angles to the field we can obtain a continuous rotation of the coil. This is done with a device known as a commutator, and the result is a simple electric motor. It is a D.C. (direct current) motor, since it can be driven from a simple battery such as an accumulator, the current reversal being achieved by the commutator itself.

If a coil such as we have described is not entirely free to rotate but is restrained by elastic springs, then the coil will come to rest when the magnetic twist on the coil is equal to the restraining twist of the springs. Obviously the greater the magnetic twist the further will the coil rotate, and since the magnetic twist is proportional to the current in the coil we can use the rotation to measure the current flowing. This is the principle of moving coil galvanometers or ammeters which are used for measuring electric currents. In the ammeter the rotation of the coil is measured by attaching a long light pointer to it and allowing it to move over a circular graduated scale.

Since magnetic fields are the same whether they are produced by permanent magnets or by currents it is to be anticipated that two electric currents will exert a force on each other. The existence of such forces was first demonstrated by Ampère who showed that two straight parallel wires carrying currents attract each other when the currents in each are flowing in the same direction, but repel when the direction of one of the currents is reversed. At first sight it may seem strange that like currents attract and unlike ones repel, but if we remember that the magnetic lines of force round a wire are circles it is an easy matter to see from Fleming's Left Hand Rule that this should be the case. This mutual attraction and repulsion between currents can be used to measure current by counterbalancing the force between two coils by a weight. The current is then measured in

Magnetism

terms of the physical dimensions of the coil and the weight required to balance the coils. These quantities value only the measurement of length, mass, and time, and so the current can be measured 'absolutely', i.e. without reference to any other magnetic or electrical quantity arbitrarily chosen as a standard. Such measurements are rare in day-to-day electrical measurements, where the usual practice is merely to compare one current or resistance with another, so it is as well to be reminded that absolute measurements in electricity can be made.

Once Oersted had shown how an electric current could produce a magnetic effect, the idea arose that a converse effect might occur, amounting to the conversion of magnetism into electricity. Reasoning that since a steady current gives rise to a steady magnetic field the converse should also be true, most experimenters merely placed a magnet (or a coil carrying a steady current) in the neighbourhood of a coil of wire and looked for a current in the wire – every time without success. Eleven years elapsed during which this false trail was exclusively followed, and it took an experimenter of Faraday's genius to discover the conditions in which the conversion of magnetism to electricity does take place. His original experiments were carried out with two coils of wire, placed near one another but without making any direct physical contact one with another. On passing a current through one of the coils a secondary current was observed to flow in the other coil, which however persisted only as long as the primary current was changing. Once the primary current was established no secondary current was observed. We would describe this occurrence today in terms of a current being induced in the secondary coil by the changing current in the primary. We use the term induction quite generally to indicate any effect produced by one body on another without their being in actual physical contact, having already employed this terminology implicitly in the statement that a magnetic field induces magnetization in iron. The essence of Faraday's discovery is that the mere existence of a primary current is not sufficient to induce a secondary current; some variation of the primary current, for example starting or stopping, is essential. Faraday soon showed that similar effects

Later Developments – Electromagnetism

could be produced by moving a magnet in the neighbourhood of a coil, or more generally by moving the coil about in the magnetic field of either a permanent magnet or a coil carrying an electric current. But as soon as the motion ceased, or in other words, as soon as the element of change was eliminated, the induced current vanished. So the conversion of magnetism into electricity requires the same factor as the production of magnetic effects by electric charges – motion or more generally the element of variation with time.

Faraday's own description of these results was in terms of the idea of lines of force. We have encountered this concept already, but it seems to have been Faraday who suggested that they may serve as something more than a mere qualitative description of things. Faraday regarded all space as being filled with these lines of force, the lines themselves being in a state of tension and thus tending to be as straight as possible. He supposed that the number of lines of force which intersect an area 1 square centimetre at right angles to the lines could be used as a measure of the strength of the magnetic field. This idea of investing semi-fictitious lines with a quantitative physical meaning was not accepted by all his contemporaries, and it is one which we ourselves should be careful of believing literally, but it cannot be denied that for Faraday it was a very fruitful concept. He showed that the current induced in a conductor is proportional to the rate at which the magnetic lines of force are cut by it. However, he went further than this by demonstrating that the act of cutting lines of force at a given rate results in the production of a definite electromotive force in the circuit; the current which flows if the circuit happens to be closed is a secondary effect. Thus it is not only necessary to move a conductor in a magnetic field in order to produce an effect; the motion must be across the lines of force. The term electromotive force introduced above is a measure of the electrical energy needed to make the charges circulate round a wire, thereby constituting an electric current. It may be regarded as a force on the charges which causes them to flow. It is invariably associated with the idea of work, and in the mechanical analogue discussed on page 19 the source of electromotive force

Magnetism

(e.m.f. for short) is the water pump, which is driven from an external agency and which maintains the pressure necessary for continuous flow of water. It is easy to appreciate that a water pump has to do work in making the water flow. If the pump is worked by hand then the arm-muscles have to provide the motive power and we tire accordingly. In a similar manner work has to be done by an electromotive force to keep the electric charges in motion and thereby maintain an electric

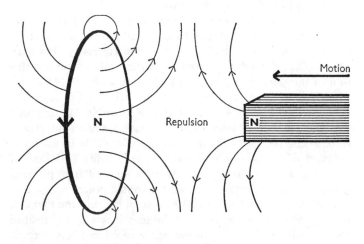

Figure 7. The induction of a current in a circular loop of wire by an approaching magnet is always such as to oppose its motion.

current. Where does the energy come from in this case? If the e.m.f. is provided by a Voltaic cell (see page 19), then it can be demonstrated that chemical reaction within it provides the energy. But we have no cells to maintain the flow of current that is induced by a changing magnetic field, and we must look elsewhere for the source of energy. Consider the induction of a current in a closed loop of wire by the passage of a bar magnet towards it as shown in Figure 7. According to Faraday the changing number of lines of force intersected by the loop as the magnet approaches gives rise to a definite e.m.f. such as to drive a current round the loop. The direction of this current is

Later Developments – Electromagnetism

given by a simple law established by Lenz in 1834. Lenz's law states that when an induced current flows it always does so in such a direction as to oppose the motion inducing the current. This is actually a special case of the more general principle of Le Chatelier, which states that when any system in equilibrium is disturbed, it always reacts in such a manner as to oppose the forces which upset the balance. Lenz's law applied to Figure 7 maintains that if the north pole of the magnet is nearest the coil, the current flows in an anti-clockwise direction as viewed from the moving magnet. Now, as we have already seen, a circular coil carrying an electric current behaves like a magnetic dipole, and the laws which govern the production of magnetic fields by currents state that in this case a south pole is formed on the left and a north pole on the right. The simplest way to remember this is by means of the statement that 'if you are looking at a clockwise current, you are looking at a south pole'. However, like poles repel, and so our moving magnet experiences a force tending to resist its motion towards the coil. In order to move it towards the coil we have to push it, and the energy associated with the induced e.m.f. comes from our own muscular effort.

This great discovery of Faraday's is usually termed electromagnetic induction, and it was his first piece of physical research, his previous activities having been in the field of chemistry. Not only was it of first-rate importance for the proper understanding of the connexion between magnetism and electricity; it brought with it the possibility of producing for the first time electricity on a large scale. For the production of electricity essentially involves changing energy from one form into another. In the voltaic cell energy of chemical formation is converted into electrical energy, but the amount of energy which can be converted directly into electrical energy by this means is really exceedingly small. In order to manufacture electricity on a large scale some means of converting mechanical energy into electricity is required, for there is a vast store of mechanical energy potentially available in the form of high altitude lakes and the tides. Moreover, whereas the direct conversion of chemical energy to electricity can only be done on a

Magnetism

tiny scale its conversion (by combustion) to mechanical energy is not limited in this way, and indeed the means of producing continuous mechanical movement was becoming readily available at the time of Faraday's discovery as a result of the development of the reciprocating steam engine. The discovery of electro-magnetic induction established the feasibility of producing electricity by mechanical means and in this sense Faraday can rightly be regarded as the father of electricity as we understand it today.

The simple electric motors invented by Faraday and described previously are reversible in the sense that the roles of the current and the motion may be interchanged. If the wheel in Figure 5 is rotated by an external force a steady e.m.f. is induced between the rim of the wheel and its axle, and this may be used to drive a current round an external circuit. Machines used in this way are called dynamos, and the two so far described are unusual in that they produce a steady unvarying e.m.f. and direct current.

The more common type of dynamo consists basically of a rectangular coil of wire rotating in a magnetic field. As the coil is rotated lines of force are cut and an e.m.f. is produced which is led out of the coil by sliding contacts. If the coil is rotated at constant speed it is easy to see that the rate of cutting lines of force will not be constant. The e.m.f. therefore is not constant but varies with the instantaneous position of the coil, i.e. with time. The variation is smooth and the e.m.f. is alternately first in one direction and then in the reverse direction every complete revolution. This is called an alternating e.m.f. and it gives rise to alternating current. The e.m.f. consists of a number of identical cycles repeated at a definite frequency of so many a second. Power stations deliver alternating current at a frequency of 50 cycles a second (50 c/s) in England, 60 c/s in the U.S.A. The fact that alternating current changes its direction fifty times a second might be thought to be a disadvantage, but for most purposes it is not. We use electricity primarily for driving machinery and for heating. It is quite a simple matter to construct an A.C. electric motor, a fact which we should

Later Developments – Electromagnetism

be able to appreciate since an A.C. dynamo in reverse will behave as a motor; and the heating effect of an electric current turns out to be the same which ever way the current is flowing, so that alternating current produces heating just like direct current.

Actually alternating currents possess one great advantage which far outweighs any disadvantages they possess. This arises from the possibility of using transformers. A transformer consists simply of two coils of wire wound on an iron yoke, the effect of the iron being to concentrate the magnetic lines of force due to one coil so that they all pass through the other. If a current is changed in the primary coil an e.m.f. is induced in the secondary which persists as long as the primary current is changing. But an alternating current is constantly changing, so if it is applied to the primary coil an alternating e.m.f. appears across the secondary. The magnitude of this e.m.f. is proportional to the ratio of the number of turns on the secondary coil to that on the primary, and so it can be made greater or less than that applied to the primary. Thus it is very easy to change the electrical pressure with A.C. by using transformers, a thing which is very difficult with D.C. (it need hardly be said that a transformer will not function with D.C.).

If D.C. is necessary for some special purpose it is always possible to convert A.C. into D.C. There are a number of ways of achieving this, but the simplest is to use a rectifier, which is a device which allows current to flow through it in one direction only.

So far we have considered only the current induced in wires and in closed loops of wire. But currents may be induced in solid conductors as well. These currents are usually termed eddy currents. Their strength and direction are given by Faraday's and Lenz's laws, so that if in Figure 7, the circular coil were replaced by a sheet of solid metal, electric current would be induced in the sheet by the approaching magnet. In this case the current would flow in circles in the plane of the sheet in such a sense as to oppose the motion of the magnet. Such currents, induced in solid conductors, are generally known as eddy currents. Eddy currents are usually a nuisance, although they can be

Magnetism

put to good use, and we shall have more to say about them later.

Suppose now the metal sheet were replaced by an insulator, such as wood; what effect does an approaching magnet have on it? Faraday himself gave no reply to this question, but it was answered some years later by a devout follower of Faraday, James Clerk-Maxwell (1831–79), whose brilliant mathematical researches ultimately led him to a final synthesis of electricity and magnetism. In order to understand Maxwell's work we shall find it necessary to extend our acquaintance with some important concepts in electricity.

We have already seen that the idea of electrical charges grew up in order to explain the force of repulsion between two pieces of rubbed amber, a force which decreases with the distance between the charges according to an inverse square law. Now we can describe the attraction between charges in terms of an electric field in just the same way as we described the force between poles in terms of a magnetic field, and we can use the idea of electric lines of force originating from positive charges and all terminating on negative charges in a similar manner. But there is one important difference. Whereas the concept of magnetic poles turned out to be merely a convenient fiction, just one way of describing the observed phenomena, no experiment or theory has cast any doubt upon the physical reality of electric charges. This does not mean that we know what electrical charges are, for we are still only describing physical phenomena in terms of a concept. What it does mean is that as soon as we admit the existence of electrical charges it no longer becomes necessary to explain magnetic effects in terms of magnetic poles, since moving charges (i.e. electric currents) can account for them just as well. Since modern atomic theory confirms that almost all the particles which constitute matter carry an electric charge it is clear that the right way of looking at magnetism is as an effect associated with electricity.

Maxwell showed that when a magnet is moved the effect of the moving lines of force is to generate an electric field everywhere in the vicinity. Now it is possible to show that the electric field

Later Developments – Electromagnetism

strength is the electrical potential difference between two points one centimetre apart, so that if a magnet is moved near a conductor a current will be induced in that conductor; it does not matter whether we speak of the current as being due to an electric potential difference, e.m.f., or electric field, since one cannot exist without the other. But when a magnet is moved with respect to an insulator Maxwell's theory maintains that the prime effect is the production of an electric field. The converse effect must also be true. Any changing electric field must inevitably result in the production of a magnetic field. A changing electric field may be produced in many ways, for example by rotating a charged body. When this was done* it was found to give rise to a magnetic field of magnitude equal to that predicted by Maxwell's theory. The reader may have noticed that this so-called converse effect is not actually new, for we have already used the idea that a moving charge constitutes an electric current, a fact realized by Faraday himself. Nevertheless it illustrates one of the cardinal features of Maxwell's theory and indeed the course which the theory of electricity has subsequently followed – the gradual elimination of charges and poles and the ultimate description of all electrical and magnetic phenomena in terms of magnetic and electric fields.

We have already mentioned how a system of electrical and magnetic units may be built up from the inverse square law for magnetic poles, namely

$$F = \frac{m_1 m_2}{\mu d^2}$$

This formula leads straight away to a definition of magnetic field strength, and we can define a unit of current in terms of the magnetic field it produces. The procedure can be extended to establish a complete electromagnetic system of electrical units. But the force between two electrical charges also obeys an inverse square law, and we can define an electrostatic unit of

* The experiment is not as simple as it sounds since the magnetic field produced is very small. It was first successfully carried out by the American physicist H. A. Rowland in 1876.

Magnetism

charge in the same way as we define an electromagnetic unit of pole strength, namely from an expression

$$F = \frac{q_1 q_2}{kd^2}$$

where q_1 and q_2 are the charges and k is a constant known as the dielectric constant of the medium, introduced to take into account the fact that the force between two charges depends on the nature of the material between them. But an electric current is simply a flow of charge and can be equally well defined as the charge which flows past a given point in one second. Thus one can proceed to build up an electrostatic system of electrical units. However, the two systems of units must be related – one can measure wealth either in pounds or in dollars but so many dollars must always be the equivalent of so many pounds – and by measuring any one electrical quantity, for example electric current, in the two systems of units the relation between them can be found. The relation turns out to be quite startling. The constants μ and k, introduced merely to make the expression for the forces between charges and poles more general, always appear in the relation between the two systems of units, not singly, but always as the product μk. For example the theory shows that one electrostatic unit of electric charge is the same amount as $\sqrt{\mu k}$ electromagnetic units of charge, and when an experiment was carried out to measure the same quantity of charge in the two systems of units, the value of $\sqrt{\mu k}$ thereby determined was found to be about $\frac{1}{3 \times 10^{10}}$ Now from the way it appeared in the mathematical equation it was evident that the quantity $1/\sqrt{\mu k}$ represented a velocity, and the electrical experiments demonstrated that $1/\sqrt{\mu k} = 3 \times 10^{10}$ centimetres per second. This is almost exactly equal to the velocity of light.

Let us pursue one of Maxwell's ideas a little further.

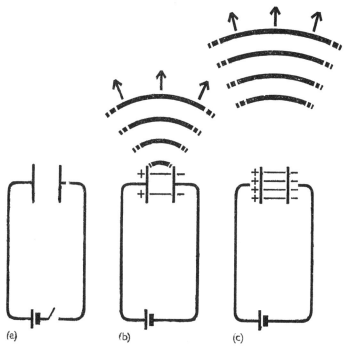

Figure 8. When an e.m.f. is applied to a condenser the charges flow along the wire and come to rest upon its plates, setting up an electric potential difference across the plates which, when the condenser is fully charged, exactly opposes the e.m.f. of the battery and stops further flow of current. While the condenser is being charged there is a flow of current in the wire and this current is transmitted across the peak by a displacement current (Maxwell's term) proportional to the rate of change of electric field between the plates. The changing electric field gives rise to a magnetic field which spreads outwards from the condenser. (a) Key open, no current. (b) Key closed, conduction current through the wire, displacement current across the condenser, and magnetic field being generated by the changing electric field. (c) Key closed, but condenser fully charged, conduction and displacement currents zero, and magnetic field travelling outwards.

Magnetism

Suppose we have two parallel conducting plates separated by an insulating medium, such as air. Such a system of insulated plates is termed a condenser. If we now connect these plates to the terminals of a battery, charges will flow and accumulate on the condenser plates until the potential difference between the plates is equal to the e.m.f. of the battery and the current ceases to flow (Figure 8). Now as the charge is building up on the plate an electric field is caused to exist between them which increases until the flow of charge ceases. According to Maxwell's theory the changing electric field between the condenser plates produces a magnetic field in their vicinity. This magnetic field is propagated with a quite definite velocity which we denote by C, and Maxwell was able to calculate what this velocity should be. He found that the velocity is given by the equation

$$C^2 = \frac{1}{\mu k}$$

If the battery in Figure 8 is replaced by an A.C. generator then the charges on the condenser plates are continually building up, dying away, and reversing their sign. In such a case the magnetic field varies in the same manner, and we can say that the magnetic field is propagated as an electro-magnetic wave with the same velocity C. The electrical measurements indicated that C was very close to the velocity of light and Maxwell inferred from this that light, which was by that time known to be a wave motion, was indeed an electromagnetic wave. More refined experiments confirmed the identity of C with the velocity of light, which is almost exactly 3×10^{10} centimetres per second. In an electromagnetic wave the magnetic field and electric fields, for these are present as well, vary with both time and distance. The time taken for a wave to repeat itself is called the period of the wave, the distance between successive crests is called the wavelength, usually denoted by the Greek letter λ, and the number of complete waves per second is termed the frequency, for which the usual symbol is another Greek letter v. For any wave the product of these two quantities equals the velocity of propagation of the wave, viz.

$$C = v\lambda$$

Later Developments – Electromagnetism

In any real electromagnetic wave the electric and magnetic fields are always at right-angles to one another and both are at right-angles to the line along which the wave is being propagated. In addition to light waves a large number of other forms of electromagnetic radiation are now known to exist. These differ from one another only in wavelength. The longest known rays are those used for long-wave wireless transmission. The wavelength of the long-wave transmitter at Droitwich for example is 150,000 cm.; it radiates at a frequency of 200,000 c/s and these two quantities when multiplied together give 3×10^{10} for their velocity as we should expect. Wireless waves of much shorter wavelength can be produced; those used for television have a wavelength of a few hundred centimetres while those used for radar purposes may be only a few centimetres. As the wavelength decreases we approach the visible light waves, but immediately before we enter this region we pass through the so-called infra-red part of the spectrum. These waves vary from one thousandth to one ten-thousandth of a centimetre and give us the physical sensation of heat. Our eyes are sensitive to only a narrow range of wavelengths, from about three to seven-hundred thousandths (3 to 7×10^{-5} cm.) of a centimetre. As the wavelength becomes shorter we pass into the ultra-violet region which provides us with our sun-tan. One of the more remarkable properties of electromagnetic waves is that as their wavelengths become shorter their energy becomes greater, and the energy of ultra-violet waves is just short enough to be dangerous to human beings in large doses. This is of course why we develop a sun-tan. It protects our bodies from absorbing too much ultra-violet light by giving us an opaque covering. As the wavelength becomes shorter the region of X-Rays and finally γ rays is reached. γ rays possess the shortest wavelengths at present known to us. They are emitted from radio-active bodies and are the rays that would do the most damage to human beings after the explosion of a nuclear bomb.

Maxwell's theory of electric and magnetic fields is embodied in a set of four mathematical equations, and from these and together with one or two other subsidiary equations (equations such as Ohm's Law, which do not tell us anything more about

Magnetism

electricity and magnetism but only about the way certain substances react to magnetic and electric fields) the whole range of magnetic and electrical effects, in which we should include light and its reflection and refraction, can be derived. The progress of science consists essentially in compressing more and more densely all the observed facts into a single all-embracing hypothesis, and in this sense electricity and magnetism can be regarded as being in a fairly advanced state of development.

CHAPTER 4

A General Synopsis of Magnetic Behaviour

WHEN we speak of a substance as being magnetic we usually imply that it has properties similar to those of iron. By this we mean that if it is placed in a magnetic field, which may be produced either by a coil of wire or by another magnet nearby, it becomes magnetized and behaves like a magnet. The question now arises as to whether in this sense any other substances can be regarded as magnetic. It would indeed be remarkable if, of all the substances known to man, only iron should be magnetic, and it is not altogether surprising that other substances exhibit this property as well. However it was virtually impossible for the early experimenters to establish this fact, for as we shall see the magnetization acquired by a body is proportional to the strength of the magnetic field in which it is situated. Most substances are so feebly magnetic that it is necessary to place them in very strong fields before any effect can be detected. But until Oersted's discovery that an electric current can give rise to a magnetic field, the only way the early experimenters had of magnetizing a substance was to place it near a lodestone, and the magnetic fields applicable in this way were far too weak to induce a detectable magnetization in any substance other than iron. Thus it was that the greater part of the early theory of magnetism, including Ampère's hypothesis of minute circulating currents as the ultimate explanation of magnetism, was developed at a time when the only substances known to be magnetic were iron and lodestone. The ideas and formulations of electromagnetism due to Laplace, Ampère, and Poisson are nevertheless as relevant today as when they were first conceived, and this remarkable achievement is rightly regarded as one of the supreme triumphs of scientific thought and imagination.

However with Oersted's discovery it became possible to make a piece of iron into a magnet far stronger than one of lodestone, merely by winding a coil of wire on it and passing a current

Magnetism

through the wire. This is the principle of the electromagnet, which has become one of the most important tools in modern physical research. The first true electromagnet seems to have been constructed by William Sturgeon (1783–1850) in 1825. It was made of soft iron bent into the shape of a horseshoe five inches high, and was wound with eighteen turns of wire. On passing a current through the wire it was found that the electromagnet could lift twenty times its own weight. This is an indication of the very large magnetic forces which can be produced by even the simplest electromagnet, and the reason is that iron is not only very strongly magnetic but it is also very easily magnetized. Some years later Joseph Henry* (1797–1878) in America constructed an electromagnet of much superior design, and the performance of Sturgeon's original electromagnet, however remarkable it had appeared to his contemporaries, was rapidly surpassed.

The truly enormous magnetic fields (fields up to several thousand oesteds were quite feasible; the chief limitation was that the batteries used to supply the current ran down rapidly) available with electromagnets made it possible for the first time to carry out, with reasonable hopes of success, experiments to find out whether substances other than iron are affected by a magnetic field.

The first to carry out a systematic investigation was Faraday, who showed that all substances are magnetic to a certain extent although in most cases the magnetic effects are very feeble indeed.

It is possible to detect the presence of magnetization in a body by observing the force on that body in a magnetic field. The magnetic field must not be uniform, for, as we have seen, no magnet experiences a force in a uniform magnetic field, and this is true irrespective of how the magnet has been made or of the material of which the magnet is composed. Fortunately it is a good deal easier to make non-uniform magnetic fields than uniform ones, so that this is no disadvantage. By careful measurement of the force experienced by a substance in a non-uniform

* Henry also discovered electromagnetic induction independently of Faraday. He is best known today for his discovery of self-induction.

A General Synopsis of Magnetic Behaviour

magnetic field Faraday was able to distinguish three classes of materials. Many substances placed in such a field are attracted slightly towards the region where the magnetic field is strongest. If we imagine the field as being due to a bar magnet these substances will be urged towards one of the poles of the magnet, since it is here that the field is most intense. These substances are termed paramagnetic. Evidently in paramagnetic substances the magnetic field due to a magnet induces a magnetization in the same direction as in the original magnet, so that a south pole is induced opposite the north pole of the magnet and there is a force of attraction between them. The force is however exceedingly small and usually corresponds to a magnetization about a million times smaller than that of iron. Typical paramagnetic substances are sodium, oxygen, and the salts of iron and nickel.

On the other hand some substances do just the opposite; they are repelled by a strong magnetic field and always try to find their way to its weakest parts. These are known as diamagnetic and their behaviour indicates that the induced magnetization is opposite to that of the field. Typical diamagnetic substances are glass, water, and mercury. Diamagnetic substances are usually very weakly magnetic indeed, less magnetic even than paramagnetics, although bismuth is one exception to this rule.

The third class of substances includes iron and these are known as ferromagnetic. Ferromagnetic substances behave like paramagnetics in that they are attracted towards the strongest part of a magnetic field, but the effect is very much greater. Moreover there are no substances with properties intermediate between those of iron and those of a typical paramagnetic,* and this lends support to the view that iron and iron-like substances are in a class of their own and are not to be regarded merely as extra-strong paramagnetics. For many thousands of years iron was the only known ferromagnetic substance, but nowadays many more are known that can legitimately be termed

* This rather sweeping statement would no longer be regarded as being strictly true, and indeed it may occasionally be difficult to decide whether a substance is truly ferromagnetic or strongly paramagnetic from its observed behaviour.

Magnetism

ferromagnetic. The elements iron, nickel, and cobalt and all their alloys are now known to be ferromagnetic, and in addition some rare-earth metals (see page 118) are ferromagnetic at lower temperatures. A large number of oxides and mixtures of oxides of the ferromagnetic metals are magnetic, and although the name ferrimagnetic is usually reserved for these they behave essentially as ferromagnetics. We shall devote a whole chapter to these interesting materials, but it is sufficient here to remark that one of these oxides of iron whose chemical formula is Fe_3O_4 is none other than the lodestone by whose properties magnetism was first discovered.

Our classification of magnetic types has been so far purely qualitative. However, within any one group of substances it is found that some are more strongly magnetic than others, and we can make only limited progress unless we can make this statement more precise. As mentioned already, the magnetism of para- and diamagnetic substances is usually determined by measuring the force exerted on them by a non-uniform magnetic field. But we cannot use this force straight away as a measure of the magnetization of a substance, for the force on one substance may be greater than that on another merely because it happens to be larger in size.

If we imagine a magnet to be made up of a large number of small circular currents then each current will produce its own lines of force, and if the circular currents are parallel the lines will continue throughout the interior of the magnet, as shown in Figure 9. These lines of force constitute a quantity known as the magnetic induction, usually given the symbol B, and are similar to the magnetic lines of force used to describe a magnetic field. The magnetic induction can now be defined as the total number of lines intersecting a unit area at right angles; we can speak of it as so many lines per square centimetre. Obviously the magnetic induction inside a magnet is much greater than at a point well outside it, since although the total number of lines is the same inside and out they are more crowded together inside. The lines of force have a definite direction and can be likened to a flow of water along a tube. For this reason the lines are sometimes referred to as magnetic flux, and the magnetic

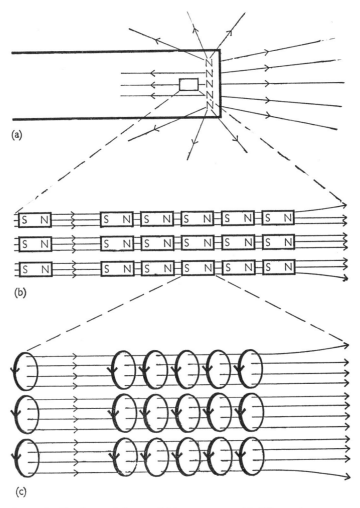

Figure 9. The lines of force within a magnet. (a) Where the magnet terminates, the lines of magnetic poles at the ends give rise to a magnetic (demagnetizing) field which, within the magnet, is in the opposite direction to that of the induction. (b) Lines of induction due to elementary magnets. (c) The same, with the elementary magnets replaced by circular molecular or atomic currents.

Magnetism

induction is often known as flux density, since it is the number of lines emerging at right angles from a unit cube. This terminology is often used by engineers, who speak of a flux density of so many lines per square centimetre or per square inch. The physicist's unit is the gauss, so named in honour of the famous German mathematician Karl Friedrich Gauss (1777–1855) who also made notable contributions to the theory of magnetism. Inside a bar magnet of the type used in school experiments the induction may be one or two thousand gauss, while the induction in a large electro-magnet is often between twenty and thirty thousand gauss.

Returning now to our magnet of Figure 9, we can replace each circular current by a dipole, and if we do this we notice that the dipoles terminate on the ends of the magnet, giving rise to a distribution of magnetic poles of opposite sign on each of the faces of the magnet. These poles give rise to a field which is in the opposite direction to the induction inside the magnet (Figure 9a). For this reason it is usually referred to as a demagnetizing field. If this field is denoted by H then the number of lines per unit area inside the magnet is $B-H$. This quantity is evidently a measure of the intrinsic magnetization possessed by the magnet since it measures the total number of lines less those due to any magnetic field. Since a magnetic field exists independently of whether it is produced by poles or by currents we are justified in using $B-H$ as a measure of the number of lines produced by the magnetization alone.

Suppose we write $B - H = J$. Now it turns out that J is related quite simply to another fundamental quantity known as the intensity of magnetization. This is defined as the magnetic moment of a substance divided by its volume. It is easily seen that in the case of a magnet of regular shape this is also equal to the pole strength on one of the faces divided by its area, since pole strength multiplied by length is magnetic moment which is also equal to intensity of magnetization times volume. If the intensity of magnetization is denoted by I then it can be shown that $J = 4\pi I$ so that

$$B - H = 4\pi I$$
$$\text{or} \quad B = H + 4\pi I$$

A General Synopsis of Magnetic Behaviour

Another way of arriving at the same result, which however avoids the logical inconsistencies of dealing in terms of both circular currents and magnetic poles, is the following. Consider a long cylinder of ferromagnetic material sufficiently long that the poles on its ends are so far away as to exert no influence except in their immediate vicinity, closely wound with a coil of wire. Inside the material there are lines of force due to the current in the wire the number of which per unit is just H, the magnetic field strength due to the coil. Now we can regard the magnetization of the material as being due to circular currents which cancel everywhere in the interior, but on the surface form a series of circles. So that in addition to the current flowing in the windings there is an effective current flowing in the same sense on the surface of the cylinder giving rise also to lines of force inside it. The number of lines of force per unit area due to this effective current comes out to be just $4\pi I$, so that the total number of lines per unit area inside the material, namely the induction B, is just $H + 4\pi I$.

The importance of the quantity B lies in the fact that our experiments measure B rather than H and I separately. Furthermore, when the mathematical theory of electricity and magnetism is formulated and electromagnetic effects are related to their cause, this cause is invariably found to involve B rather than I. The fact that physicists, particularly those interested in the magnetic properties of matter, prefer to separate the intrinsic magnetic effects of a substance from those of the field in which it is placed is understandable enough. But it appears from both theory and experiment that inanimate objects, such as coils of wire, are unaware of any difference between lines of forces, no matter where they originate. In Chapter 6 we shall see that all magnetic effects whether from a coil of wire or a permanent magnet are electrical in origin, so that in adding H to $4\pi I$ we are in fact adding together two quantities which originate in the same way, and in consequence we can more readily appreciate that their effects cannot be distinguished, even in principle, by experiment.

The occurrence of the factor 4π may be puzzling, but need not worry us. It comes from the system of units employed by

Magnetism

physicists, and another system of units could easily be chosen in which J is equal to I. This would mean that factors of 4π would appear in other equations, for example the law of force between magnetic poles would have to be $F = \dfrac{m_1 \times m_2}{4\pi\mu d^2}$. Such an equation would amount to a re-definition of pole strength which we are quite at liberty to make provided we arrive at a self-consistent system of units. What this really means is that we cannot avoid 4π in all our formulae, but we have considerable choice as to which formulae it appears in. Most electrical engineers prefer to put in the 4π in the early fundamental equation so as to eliminate it later on, while the majority of physicists prefer to state their basic formulae in the simplest possible way and put up with the 4π in their final equations. The important thing is that in either J or I we have a quantity which is a direct measure of the amount of magnetization possessed by a body. We shall use I hereafter. In most paramagnetics and for diamagnetic substances the ratio of the induction B to the magnetic field producing it, H, is a constant and is called the magnetic permeability μ. It is actually the same μ as appears in the law of force between two magnetic poles although it may take some ingenuity to prove this. If we take the expression $B = H + 4\pi I$ and divide throughout by H we get

$$\mu = 1 + 4\pi I/H$$
$$= 1 + 4\pi \kappa$$

The quantity κ (Greek kappa) is termed the magnetic susceptibility. It is the intensity of magnetization induced by unit field. We note that since I is the magnetic moment per unit volume κ refers also to unit volume and for this reason is often called the volume susceptibility. It is sometimes more convenient to work in terms of a magnetic moment per unit mass, and in this case a mass susceptibility can be defined in a similar manner. The mass susceptibility of any substance is equal to κ divided by the density of the substance. A rather more significant quantity, useful in discussing the magnetic properties of individual atoms and molecules, is the atomic or molecular susceptibility, equal to κ divided by the number of atoms or molecules in unit volume. κH is then the magnetic moment per atom or per

A General Synopsis of Magnetic Behaviour

molecule. Unless otherwise stated we shall hereafter use magnetic susceptibility to refer to the volume susceptibility.

We are now in a position to put our formal classification of magnetic behaviour on a proper quantitative basis. It does not much matter whether our description is in terms of μ or κ, since these are simply related, but in the case of weakly magnetic substances it is usually more convenient to work in terms of κ.

For paramagnetic substances κ is positive and usually of the order 10^{-5}. Unless the temperature is very low or the fields used for measurement are very high κ is independent of the field applied. The paramagnetic susceptibility of most substances decreases with increasing temperature and some paramagnetic substances, but by no means all, obey the so-called Curie Law (in honour of Pierre Curie, better known for his collaboration with his wife Marie Curie in studying radioactivity) which states that the susceptibility varies inversely as the absolute temperature*. In symbols

$$\kappa = \frac{C}{T}$$

in which C is a constant. This equation implies that the susceptibility should become very large at low temperatures and this indeed happens to be the case. More often paramagnetic substances obey a modified Curie Law known as the Curie-Weiss Law namely

$$\kappa = \frac{C}{T - \theta}$$

where T is the temperature of the substance in the absolute scale and θ is a constant. One might well ask what happens when T is equal to or less than θ since this would imply infinite susceptibility or the presence of magnetization without the necessity of an external field. We shall avoid this issue for the time with the statement that usually the law ceases to be obeyed before this temperature is reached, which, however frustrating, is usually true. We shall, however, return to the question in Chapter 7.

* Absolute zero is $-273°$ centigrade and the absolute scale of temperature may be obtained from the centigrade scale by adding $273°$, so that absolute zero is $0°K$, and for example the freezing point of water is $273°K$. The symbol K stands for Kelvin, who first conceived the idea.

Magnetism

Diamagnetic behaviour may be described by a negative volume susceptibility. Usually κ is smaller in magnitude than that of paramagnetics and is very often almost independent of temperature. Since κ is negative it follows that μ is less than one, and this means that the lines of force instead of crowding into a diamagnetic substance are actually squeezed out of it.

In ferromagnetic substances κ is very large – it may be as high as one million. It also depends upon the value of the magnetic field applied. It therefore has meaning only if the field applied to it is specified as well, and what is usually done in practice is to quote an entire magnetization curve, i.e. a graph showing how the induced magnetization varies with the magnetizing field. This curve exhibits certain remarkable characteristics, which are illustrated in Figure 10. When a small magnetic field is applied the magnetization increases slowly at first and is approximately proportional to the field. As long as this proportionality is maintained one can take the ratio of the induction to the field as defining a quantity characteristic of the substance. It is called the initial susceptibility and is a very important criterion of suitability for certain applications.

As the field is increased the magnetization curve becomes steeper and a new phenomenon now appears, for if this field is removed the substance retains some of its magnetism. The amount retained is termed the remanence and it is this feature of the magnetization of ferromagnetic substance that enables them to be made into permanent magnets. If the field is increased further the curve flattens out, and a stage is reached at which no further increase in magnetization takes place no matter how large a field is applied.*

When this stage has been reached the substance is said to be magnetically saturated. If the field is now removed the material

* Actually there is a slight increase of magnetization with field after saturation has been reached. Usually the magnetization increases by no more than about 1 gauss for every 10,000 oersteds for iron and nickel at room temperature, so this effect may usually be ignored in the discussion of magnetization curves.

A General Synopsis of Magnetic Behaviour

is left with its maximum remanent magnetization, and in order to reduce this magnetization to zero it is now necessary to apply a field in the reverse direction. Only then does the magnetization decrease, eventually to zero, and this occurs when the reverse field attains a definite strength; it is known as the coercive field or coercive force. If the field is now increased, still in the reverse direction, beyond the coercive force, the magnetization begins to increase once more and finally reaches saturation in the reverse direction, the value of the saturation magnetization in the two directions being identical. Saturation in the original direction can be regained by repetition of the

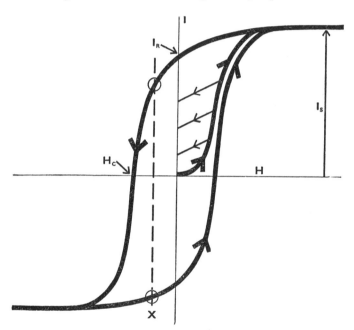

Figure 10. Complete hysteresis cycle of a typical ferromagnetic material. I_R, remanent magnetization or remanence; H_C, coercive force; I_S, saturation magnetization.

Magnetism

whole sequence of operations, in which case the magnetization curve becomes a closed loop, as shown in Figure 10. One of the features of this loop is that the magnetization acquired by a ferromagnetic substance is not a unique function of the applied magnetic field but depends upon its previous history. For example a ferromagnetic substance in a field represented by X on Figure 10 may have two quite different magnetizations depending on whether it was previously at a higher or lower magnetization. This phenomenon is termed hysteresis. Hysteresis and remanence are by no means confined to the magnetization of iron and similar materials. Many substances including metals and rubber exhibit both effects when deformed by bending, twisting, or stretching. The remanent stretch of rubber for example is easily observed in a rubber balloon when the air is let out after it has once been inflated.

A complete, closed, magnetization curve is known as a hysteresis cycle, and the area enclosed by such a closed curve is a measure of the work done by the applied magnetic field in magnetizing the substance. It is quite easy to measure the area enclosed by plotting the magnetization curve on squared paper and counting the number of squares inside. The work done in magnetizing the substance is absorbed by it from the magnetizing field and appears as heat. The amount of heat absorbed by one cubic centimetre over a complete hysteresis cycle is quite small, usually about 5,000 ergs* for iron, sufficient to raise its temperature by about $10^{-4}°C$. However this represents wasted power and we must bear in mind that in most practical applications the iron is likely to be used in A.C. operation where it is taken through a hysteresis cycle 50 times each second. Everything is done therefore to reduce the hysteresis loss in materials which are likely to be used in these circumstances. The hysteresis loss itself is determined largely by the remanence and the coercive force of the material. The coercive force and indeed the general shape of the whole magnetization curve are found to be very

* The erg is the unit of energy on the c.g.s. system. It is a small unit, useful to the physicist but too small for practical purposes where it is replaced by the Joule, equal to 10^7 ergs. A rate of working of one Joule per second is one Watt and there are 746 watts in one horse-power. A one-kilowatt electric fire consumes energy at the rate of 10^{10} ergs per second.

A General Synopsis of Magnetic Behaviour

sensitive to the mechanical state of the material. For example a piece of pure soft iron may have a steep magnetization curve and a coercive force of about half an oersted. If this same piece is mechanically worked, for example by bending or by hammering, the magnetization curve will become much less steep and the coercive force may be increased tenfold. The saturation magnetization, however, remains unchanged by this treatment. Evidently the coercive force is in some way sensitive to the internal structure of the material while the saturation magnetization is structure-insensitive. In fact the saturation magnetization proves to depend only upon the nature of the material and its temperature, the values for iron and nickel at room temperature being 1,720 and 500 gauss respectively. The remanence of a substance is often fairly insensitive to mechanical handling and usually is about one half the saturation magnetization.

We can now appreciate the tremendous technical importance of ferromagnetic materials in general and of iron in particular, for it is a very remarkable thing that not only does iron possess very nearly the largest saturation magnetization of any known substance, it is in addition by far the cheapest and most abundant of the ferromagnetic elements. All forms of electrical machinery which work on the principle of electromagnetic induction, and this includes not only dynamos and motors but transformers as well, make extensive use of iron in some form or other to increase their efficiency. Now we have seen that Faraday's Laws state that the e.m.f. induced in a conductor due to changing lines of force is proportional to the rate at which the number of lines are intersected by it, or using the terminology developed at the beginning of this chapter, to the rate of change of magnetic flux. But the magnetic flux is determined by the magnetic induction, and this is equal to the product of the magnetic permeability of a substance and the field in which it is placed, so that if a piece of iron having a permeability of 1000 is placed inside a coil of wire the number of lines, i.e. the flux, is increased by one thousand. If the current in this coil is made to change, the e.m.f. which will be set up in any nearby conduction is increased one thousand times by the presence of

Magnetism

the iron within the coil. Of course in this case some energy is wasted in the iron as heat as a result of hysteresis, but this is more than compensated by the increased e.m.f. induced in the secondary coil. Evidently for this type of application a material is needed which has a high permeability accompanied by a low hysteresis loss, and this is usually achieved by adding a small amount of the element silicon to iron. We shall see the reason for this later, but it is quoted here as an example of the way in which magnetic materials can be 'tailored' to suit the requirements of any one particular application.

Another important application of the use of iron is in the electromagnet. Sturgeon's first electromagnet was able to lift so much iron because the attracting power of a magnet is proportional to the external field it produces, and both this, and its lifting power when the objects are in contact with the magnet, are determined by the induction in the magnet, not solely by the field produced by the windings on it. For this purpose hysteresis hardly matters at all: what is needed is the largest possible induction in the magnet. This is done, partly by careful design of the shape of the iron yoke and the position of the windings, but also by adding cobalt to iron, which increases its saturation magnetization slightly.

We have already seen that one of the distinguishing features of para- and dia-magnetic substances is their characteristic variation of behaviour with temperature. How do the magnetic properties of a ferromagnetic material vary with temperature? As a result of the remarkable diversity of behaviour of ferromagnetic materials we must not expect too definite an answer to this question, but as far as the general features of ferromagnetics are concerned the answer is simple enough. All ferromagnetic substances lose their ferromagnetism at a certain characteristic temperature; iron at 770° C, nickel at 360° C. The temperature at which ferromagnetism disappears is known as the Curie temperature, and a ferromagnetic substance above its Curie temperature behaves just like an ordinary paramagnetic. This vanishing of ferromagnetism takes place quite rapidly (see Figure 11) although not as abruptly as the vanishing of the mechanical properties of a solid when it melts (ice to water

CHAPTER 5

The Atomic Theory of Matter

ONE of the tasks of the research worker in magnetism is to interpret the magnetic behaviour of substances in the light of his knowledge of their constituents and what is known about their structure, and it is impossible to proceed very much beyond the formal classification of magnetic behaviour into its three 'isms' unless one has an idea about what these constituents are likely to be, and how they behave when they are welded together to form matter.

The idea that all matter is ultimately formed of atoms is quite old and dates back in one form or other to the Greeks, but modern atomic theory is usually associated with the name of John Dalton, an English chemist from Cumberland. From a chemical standpoint the simplest substances are the elements, because these contain atoms of only one kind. One hundred and three different chemical elements are now known, although twelve of these do not occur naturally and are the product of nuclear physics rather than the chemical laboratory. Atoms may exist separately or in pairs, or in chemical combination with different atoms to form molecules, and when two elements combine to form a chemical compound they always do so in definite proportions by weight. One of the simplest examples is the combination of hydrogen and oxygen to form water, for this always occurs such that the two are combined in the ratio of one to eight by weight respectively. If nine grams of oxygen are mixed with one gram of hydrogen, and chemical combination is allowed to take place, it is found that one gram of hydrogen combines with eight grams of oxygen to form nine grams of water, leaving one gram of oxygen chemically unchanged. The same is true of all chemical reactions, and it is always found that the combining weights of two elements are in the ratio of two whole numbers. This strongly suggests that the actual weight of an individual atom is characteristic, and that the molecule of water

A General Synopsis of Magnetic Behaviour

magnetize it. This is rather less simple than it sounds. One might think that it would be necessary to apply a reverse field equal to the coercive force, thus reducing the magnetization to zero, and remove it. Unfortunately the state of zero magnetization at the coercive force is not a stable one, and if the field is removed the magnetization reverts to a value only a little less than that which it possessed at remanence. The only way to achieve complete and permanent demagnetization is to take the substance round a number of hysteresis loops with a field of decreasing maximum value. In practice this can be accomplished by using an alternating field produced by A.C. in a coil placed east-west so that the earth's field does not act upon the object to be demagnetized, and reducing the amplitude of the field to zero by gradual diminution of the current. Small screwdrivers and other steel tools which have become magnetized can usually be demagnetized in this way.

The description of magnetic behaviour given in this chapter leads us to a broad classification into three types, and our next task is to explain, as far as we are able, the observed differences in behaviour exhibited by various substances. If we are intrigued by the fact that hydrogen is diamagnetic whereas oxygen is paramagnetic, or that iron is roughly three times as strongly ferromagnetic as nickel, and wish to find an explanation for these and other observations, then our first task must be to enquire into the present-day physicists' view of matter, namely what it is made of, and how the parts fit together. Then we may stand some chance of understanding how and why it behaves as it does.

Magnetism

changes, although small, can usually be detected and, with modern techniques, measured with fair accuracy. The most important of these is the change of length that occurs when a substance is magnetized. The effect is termed magnetostriction. Even in ferromagnetics the effect is small, usually a few parts per million, but it is of great importance not only because it proves to be another of these structure insensitive properties which, incidentally, largely determines the initial susceptibility of a ferromagnetic substance, but also because it provides us with a means of converting electrical energy into mechanical energy. For example a bar of iron subject to an alternating magnetic field undergoes similar alternating extensions and contractions: this is one of the reasons why large power transformers emit an audible hum when in operation. More usefully we can utilitize a vibrating iron rod set into oscillation in this manner, as a convenient source of very high frequency sound waves (see Chapter 12). Other physical properties which change with magnetization are electrical resistance and certain mechanical properties, e.g. Young's Modulus of elasticity. For most materials these changes usually amount to less than one part in a hundred and with one notable exception are without much consequence. The one exception is the result of a magnetic field on a pocket or wrist watch. Most of us are aware that a watch is usually damaged by placing it near a strong magnet. It may stop altogether, and even if this does not happen its timekeeping powers are nearly always impaired. It may not always be possible to give a single reason for this, since a watch is a complicated mechanism usually containing a large amount of steel in its construction. If the hair spring becomes magnetized however, each part becomes like a small magnet and attracts another part. If this attraction is great enough the spring will be seriously distorted from its spiral shape and may touch in places, causing the watch to stop. But even if it is not magnetized strongly enough to do this it will no longer keep correct time, because the hair spring controls the period of oscillation of the balance wheel not only by its length but by its elasticity and this becomes altered if it is magnetized. We can only hope to restore the watch to its former state by attempting to de-

A General Synopsis of Magnetic Behaviour

for example). Moreover the disappearance of ferromagnetism is not usually associated with any large structural changes, as in the transition from solid to liquid. We shall have more to say about this later, and for the time it is sufficient to observe that the Curie temperature, like the saturation magnetization, is an intrinsic property of a substance, unaffected by mechanical means. These two are in fact the fundamental ferromagnetic characteristics: the ease with which a substance may be magnetized, i.e. the permeability, however technically important, should rightly be regarded as a secondary feature.

The effects of a magnetic field on a ferromagnetic substance are so great that it may reasonably be asked whether physical properties other than purely magnetic ones are affected. In fact it is found that all physical properties of all substances are

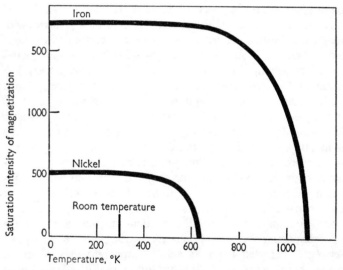

Figure 11. The variation of the saturation magnetization of iron and nickel with absolute temperature.

changed slightly by the presence of a magnetic field, though in para- and diamagnetics the changes may be so small as to be virtually undetectable. In ferromagnetics on the other hand the

The Atomic Theory of Matter

is made up either of one atom of hydrogen and one atom of oxygen weighing eight times as much, or two atoms of hydrogen and one atom of oxygen weighing sixteen times as much. It turns out that in order to explain the chemical combination of hydrogen and oxygen with other elements the latter explanation is the one which must be preferred. The essential point is that by accurately determining just how many grams of a substance combine with so much hydrogen or oxygen one can eventually reach a consistent classification of all atoms into their respective weights; one can in fact determine the ratio of the weight of any atom to that of hydrogen. This ratio is termed the atomic weight of a substance. If we take the atomic weight of hydrogen as 1, then the atomic weight of oxygen is 16, while that of the heaviest natural element is about 240. In addition to atomic weights one can extend the system to molecular weights; the molecular weight of water for example is 18 on this basis. Hydrogen gas consists of molecules each containing two atoms of hydrogen in a state of chemical combination: thus the molecular weight of hydrogen is 2, while the complex molecules of certain proteins may have molecular weights of over a million.

The next question which arises is whether the atoms have any characteristic structure of their own or differ from one another only by weight. The answer to this can readily be given, for it will be readily appreciated that an assembly of electrically neutral lumps of matter would not exhibit any magnetic properties whatever. There were several links in the chain of discoveries that led to the elucidation of the structure of atoms, notably the observation that aqueous solutions of certain salts would conduct an electric current and that air would do likewise, though to a lesser extent, if it was subjected to certain radiations. But pride of place amongst the physicists whose discoveries led to our knowledge of the structure of the atom must go to Sir J. J. Thomson (1856–1940), the son of a Manchester bookseller who went up to Cambridge in 1876 and remained there first as Cavendish Professor and finally as Master of Trinity College until his death in 1940.

A gas, such as air, under normal conditions of temperature and pressure is a very poor conductor of electricity. If, however,

Magnetism

the pressure is lowered the gas loses its insulating properties, conducts electricity, and at the same time emits a coloured light which is characteristic of the nature of the gas. The neon lighting used for nocturnal illumination makes use of this fact. If the pressure of the gas is reduced still further the luminescence disappears except for a faint fluorescence, which this time depends only on the nature of the vessel used to contain the residual gas; but the electric current through the vessel remains. The same takes place whatever the nature of the gas originally present, from which it follows that whatever is conducting the electricity under these conditions is the same whatever its origin, and must therefore be present, at least in all gases, possibly in other substances as well. By a series of ingenious experiments described in more detail in Chapter 12, J. J. Thomson succeeded in showing that the current was being carried by a rapid stream of particles each carrying a small electric charge, the mass of each particle being only 1/1840 of the mass of the lightest known element, hydrogen. These particles were subsequently found to be emitted from atoms in different circumstances, for example by radioactive substances. The same light particles are emitted when metals are exposed to ultra-violet radiation and when they are heated to a high temperature; this latter is the thermionic effect, which is the basis of the thermionic valve used in radio and television receivers. Thus it is seen that these light particles are emitted from a great diversity of materials, and it is concluded that they are a basic constituent of all matter. The name electron is given to these particles. The charge carried by an electron can be measured, and this was done very accurately by the American physicist R. A. Millikan (1868–1954), who first clearly demonstrated the fact that all electric charges are made up of one, two, or some whole number of electron charges. This discovery of the electron was of supreme importance, for not only did it point to the very close association of electricity and matter, it gave indications of the atomic nature of electricity, just as the determination of chemical atomic weight indicated the atomic nature of matter. The charge carried by an electron is negative and of magnitude $4 \cdot 8 \times 10^{-10}$ electrostatic units (e.s.u. for

The Atomic Theory of Matter

short). This means that a current of one ampere is equivalent to a flow of about 6×10^{20} electrons per second, and more sophisticated experiments have shown that the current which flows in a metal wire is indeed carried by these electrons.

If electrons are basic constituents of all atoms, how many electrons does an atom contain? As one might expect, this depends on the type of atom, but usually the number of electrons an atom contains is roughly half its atomic weight. More precisely the number of electrons is exactly equal to the quantity the chemists call the atomic number. This is a number which indicates the position of each element in the periodic table of elements and varies from one for hydrogen to 92 for uranium.

Now atoms are electrically neutral, as inferred from the fact that they are not set in motion by an electric field.* This means that part of the atom must contain a positive charge. The way in which the positive and negative charges must be arranged in an atom was first discovered by Lord Rutherford (1871–1937), perhaps the greatest experimental physicist of the twentieth century, who may properly be described as the father of nuclear energy. The detailed mathematical theory was worked out in Rutherford's laboratory by Niels Bohr (b. 1885), a young Danish physicist from Copenhagen, who was much later in life to be closely associated with the explanation of nuclear fission of uranium and the development of the atomic bomb.

The simplest atom is that of hydrogen; it is the atom of the lightest known element and contains just one electron. The hydrogen atom as envisaged by Rutherford and Bohr consists of a central massive core carrying a positive charge with the electron rotating in some form of orbit around it, rather in the way that the earth rotates round the sun. The mass of the electron is only 1/1840 of the mass of the hydrogen atom, and so nearly all the mass of the atom must be concentrated in this central core, which is known as the nucleus. Its charge must be exactly equal in magnitude but opposite in sign to that carried by the electron in order that the whole atom be electrically

---

* There may be a tendency for an atom to orientate itself in an electric field, because a large number of atoms possess electric dipole moments. The total charge carried by an atom is, however, always zero.

Magnetism

neutral. The idea that the electron actually revolves around the nucleus is necessary to explain the stability of the system, for otherwise the two charged bodies would attract each other and the electron and nucleus would coalesce. Stability can only be attained by introducing the idea of motion, so that the electrical force of attraction between the two particles is exactly balanced by the centrifugal force tending to drive the particles away from each other. On this picture the size of the atom is the radius of the orbit of the electron. This radius can be calculated using the principles of the quantum theory as we shall show later, but for the moment it is sufficient to state that for hydrogen the radius comes out to be about 10^{-8} centimetres, which is roughly what we would expect from our knowledge of the way gases behave. The nucleus on the other hand is very much smaller, only one hundred thousandth of the size of the atom, or about 10^{-13} centimetres in radius, so that the greater part of the atom is empty space. Within the atom, and of course to a lesser extent outside it, there must exist exceedingly strong electric and magnetic fields due to the charges and their motion.

So far we have not said anything about the masses of individual atoms, and chemical atomic weights merely give the ratio of the mass of an atom to that of the hydrogen atom. But the masses of individual atoms can be measured, and indeed with considerable precision, by a method which is described in Chapter 12, and which characteristically enough makes use of a powerful electromagnet. The mass of the hydrogen nucleus, which is important enough to be given a name of its own (it is called the proton) is $1 \cdot 6724 \times 10^{-24}$ gm.; that of the electron is $9 \cdot 1066 \times 10^{-28}$ gm. It is exceedingly difficult to understand the meaning of these minute figures which emphasize the extreme smallness of atoms. One way of looking at them is to realize that 1 gram of hydrogen which would occupy 22·4 litres* or slightly less than one cubic foot at ordinary atmospheric temperature and pressure would contain 6×10^{23} hydrogen atoms. (The conditional mood is necessary here because hydrogen gas is actually formed of diatomic molecules.) Other

* A litre is the metric unit of capacity approximately equal to one-and-three-quarter pints.

The Atomic Theory of Matter

atoms are heavier than hydrogen in the ratio of their chemical atomic weights and are somewhat, although not very much, larger in size. In a solid, as we shall see later, the atoms are packed much closer together than in a gas, and a one-inch cube of iron contains about 5×10^{23} atoms. In the heavier atoms there are of course more electrons, and each of these is to be regarded as revolving in its own orbit largely independent of the others. Even in a sphere as small as that taken up by a single atom there is plenty of room for them to do this, and even heavy atoms containing a large number of electrons must be regarded as having a very open structure – mostly space occupied only by intense electromagnetic fields.

Now we saw in Chapter 3 that a charged particle in motion is equivalent to an electric current, so that each of the electrons revolving in a circular orbit is equivalent to a circular current and must therefore be associated with a magnetic moment. Thus the model of the atom proposed by Rutherford and Bohr also predicts just these circular currents, the existence of which had been so brilliantly anticipated by Ampère himself almost one hundred years earlier. We can calculate the magnetic moment of an electron in its orbit, the so-called orbital magnetic moment, if we know something about the size of the orbit. In order to do this we have to invoke the assistance of quantum theory, although only in its simplest form.

One of the most striking features of present-day physics is its insistence upon the discreteness of a physical quantity. Matter is not continuous, but made up of atoms. Perhaps even more striking is the fact that electricity is always made up of a number of charges each equal to the charge on the electron. This has led to the idea of fundamental units just in the way that the charge on the electron now appears to be a natural unit by which to measure a quantity of electricity, since charges less than this do not appear to exist. Quantities which exist only as some whole number of some basic unit are said to be discrete, and many physicists believe that even such quantities as length and time may be described in this sense. One simple (and not entirely correct) way of formulating the quantum theory is by the statement that the angular momentum of an orbit is a

Magnetism

discrete quantity. The idea of angular momentum is very important in magnetism, since a charged particle produces a magnetic moment only by value of its angular momentum. This is equal to mass × velocity × radius of orbit. The unit of angular momentum is $h/2\pi$ where h is a constant known as Planck's constant of action, equal to $6\cdot27 \times 10^{-27}$ erg-seconds. A rotating body can possess angular momentum of one or two or some whole number of these units, but not values intermediate between them. It will be seen that on account of the very small size of h an ordinary body of the size we are accustomed to in everyday life can take on a range of angular momentum values which is to all intents and purposes continuous. For example a mass of 1 gram rotating in a circle of radius 1 centimetre at one revolution per second would have an angular momentum of 2 erg-seconds or roughly 6×10^{27} angular momentum units. Changing this to 6×10^{27} plus 1 would make a difference that would be completely undetectable and so the essentially discrete nature of angular moment would go unnoticed. For the electron however, with its very small mass, the situation is quite different, for an electron in an orbit may often have only a single unit of angular momentum and only rarely as many as three or four. The size of the orbit which is governed by the angular momentum of the electron is thus determined by quantum restrictions and can have only certain values, in contrast to the orbit of, for example, an artificial satellite for which the number of orbits is unlimited.*

Let us consider first the hydrogen atom pictured as a central nucleus with a single electron rotating in a circular orbit with the nucleus as centre. In order that the system be stable the electrical force of attraction between the two particles, which is equal to e^2/r^2 if r is the radius of the orbit, must equal the outward centrifugal force experienced by the electron, which is equal to mv^2/r where m is the mass of the electron and v its velocity.

Thus $$e^2/r^2 = mv^2/r$$

* The actual number is still of course limited by the finiteness of h, but for the same reasons as those given above this restriction is completely negligible.

The Atomic Theory of Matter

Furthermore the quantum theory restricts the orbit to those in which the angular momentum is $nh/2\pi$ where n is a whole number. The angular momentum is mvr and so

$$mvr = nh/2\pi$$

From this equation and the one above it is only a short step to show that the radius of the orbit r is

$$r = \frac{n^2h^2}{4\pi^2 me^2}$$

which increases as the square of the whole numbers from one upwards. The atom spends most of its time in the state corresponding to $n = 1$. This state is that of minimum angular momentum and minimum energy, and is usually referred to as the ground state. Other states corresponding to values of n greater than one exist. They have more energy and are known as excited states. In order to revert to the ground state an atom in an excited state has to get rid of its energy and it does so by emitting light. The wavelength of the light emitted as a result of the transition is governed entirely by the difference in the energies of the two states. For example, the yellow light emitted by the sodium lamps often used for street lighting is due almost entirely to electrons involved in a transition between two states of the sodium atom from $n = 2$ to $n = 1$, and although our concern is primarily with magnetic, rather than optical, effects this digression may help to illustrate one very important point, namely the close connexion between magnetism and light, a matter we shall have to return to later.

The revolving electron, being charged, is equivalent to an electric current of magnitude equal to the charge divided by the time for one complete revolution of the electron. This time is equal to $\frac{2\pi r}{v}$ if r is the radius of the orbit and v is the velocity of the electron as before. Thus the current $i = \frac{ev}{2\pi r}$. Now as shown on page 31 a current i in a circle whose area is A is equivalent to a magnetic dipole of moment iA. Therefore

Magnetism

$$\text{magnetic moment} = iA = \frac{ev}{2\pi r} A$$

$$= \frac{evr}{2}$$

since $A = \pi r^2$. If we now remember that the least angular momentum that the electron may possess is

$$mvr = \frac{h}{2\pi}$$

when $n = 1$, then the magnetic moment, usually given the symbol μ and not to be confused with magnetic permeability, is

$$\mu = \frac{evr}{2}$$

$$= \frac{eh}{4\pi m}$$

This is the smallest magnetic moment which an orbital electron may possess and seems to provide a natural unit of magnetic moment. It is usually called the Bohr magneton, and the magnetic moment of any orbital electron should be a whole number of Bohr magnetons. We notice that it depends only on h and the ratio of the charge to the mass of the electron. This ratio can be measured very accurately; it equals $1 \cdot 76 \times 10^7$ electromagnetic units of charge per gram. Putting in this value, together with that already given for h, in this formula we find $\mu = 9 \cdot 25 \times 10^{-21}$. This is the Bohr magneton, the orbital magnetic moment of the hydrogen atom in its ground state.

Heavier atoms contain more than one electron, and as we proceed from hydrogen through the periodic table of elements the number of electrons increases one by one; helium has two, lithium three, and so on up to uranium which has ninety-two. One of the predictions of the quantum theory is that the electron orbits are confined within shells, each shell corresponding to a certain value of n. The number of electrons which can be accommodated in a single shell is limited and itself depends upon n. For $n = 1$ only two electrons are required to fill the shell; for $n = 2$ there is another shell further out from the nucleus which can accommodate 8 electrons. As n increases the number of electrons which the shell can contain increases and for $n = 3, 4, 5$, the permitted number of electrons is 18, 36,

The Atomic Theory of Matter

and 50, respectively. This explains the chief features of the periodic table of elements, for the chemical properties of an element are determined very largely by the number of electrons in the outermost shell of its atom. Thus lithium, having three electrons and just one of these in the outer shell, has properties very similar to sodium, of whose eleven electrons two fill the innermost shell and eight fill the next, leaving just one in the outer shell.

Within each shell the electrons rotate in orbits and each rotating electron gives rise to a magnetic moment. These magnetic moments are, for reasons which we cannot discuss here, orientated in different directions, and the resultant magnetic moment of the atom is obtained from the vector sum of the magnetic moments of each orbital electron. When this is done the resultant effect is usually a magnetic moment of only a few Bohr magnetons even for atoms with many electrons. This is due to the tendency of the magnetic moments to point in opposite directions, so that their combined effect is zero, leaving only a few uncompensated moments to add up to the total. Actually this tendency for the orbital magnetic moments to cancel each other is so great that often the atom has no resultant magnetic moment. As we shall see later this is the situation in diamagnetic substances.

However if magnetic effects were solely due to the orbital motion of electrons magnetism would be a duller subject than it is, and in fact ferromagnetism as we know it would not exist. For it turns out that the electron possesses a magnetic moment of its own quite independent of whether it is rotating in an orbit or not. This is said to be due to electron spin, and the idea of electron spin was introduced not to explain magnetic properties but to account for certain discrepancies between the way in which light is observed to be emitted by atoms and the way theory predicted it should. Although we must not regard the electron as a rigid body capable of rotation about an axis the idea that the electron is spinning about an axis as well as rotating in an orbit is one that is often used to describe the idea of electron spin on account of the very close analogy between such a system and the earth, which not only revolves round the sun but is itself spinning on its own axis. We must not get the

Magnetism

impression that electron spin is merely an extra quantity which has to be introduced into the theory in order to make theory fit the fact. To be sure, this is the way in which the necessity of electron spin was first introduced; but the object of physics is to reduce the number of hypotheses required for the foundation of a satisfactory theory, not to increase them, and it was a great achievement for Dirac who in 1926 showed for the first time that the idea of electron spin was a necessary consequence of the electron being subject not only to the rules of quantum theory but to the requirements of the theory of relativity as well. This wonderful synthesis of quantum theory, which deals with atoms and their constituent particles, and relativity, whose chief success had so far lain in the realm of objects of astronomical dimensions, made by Dirac, remains one of the outstanding achievements of twentieth-century physics.

For us, the most important thing is that every electron in the atom possesses not only an orbital magnetic moment but a spin magnetic moment as well, and the resultant magnetic moment of the atom is the vector sum of all these. The direction of the individual magnetic moment within the atom is controlled almost entirely by the very strong electrical forces within it, and so they are not individually influenced by a magnetic field which acts only on the resultant moment. An alternative statement of the same fact is that the net magnetic moment of an atom is unaffected by an external magnetic field.

When we come to enquire how an atom possessing a magnetic moment behaves in a magnetic field the quantum theory predicts some very surprising results. A freely suspended bar magnet, it will be remembered, always sets itself with its axis along the magnetic lines of force, but it is always possible, by twisting it and holding it in position, to make the magnet point in any direction. It is not so with atomic magnets, which according to the quantum theory can take up certain orientations only, the number of orientations depending on the value of the magnetic moment. For example, if the atomic moment is one Bohr magneton only two orientations are permitted, one with the dipole lying parallel to the field and the other with the dipole in the opposite direction. If the magnetic moment is two Bohr

The Atomic Theory of Matter

magnetons there are five permitted orientations. This fact made it possible to confirm the existence of atomic magnets in a remarkably ingenious experiment due to the German physicists O. Stern and W. Gerlach (see Figure 12). They used potassium atoms for their experiment, the magnetism of which is due almost entirely to the spin of a single electron, all the effects due to the other electrons, both orbital and spin effects, having cancelled out. Now a single electron has a spin magnetic moment of one Bohr magneton and so the potassium atom as a whole can

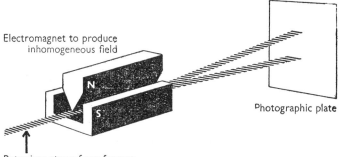

Figure 12. The experiment of Stern and Gerlach. Potassium atoms from a furnace are sent through a strongly inhomogeneous field (much stronger near the edge of the north pole than at the trough-shaped south pole). This field separates the beam into two (for explanation see text).

have only two permitted orientations in a magnetic field. Stern and Gerlach obtained their potassium atoms by heating the metal in an evacuated furnace and then passed them through a magnetic field produced by an electromagnet whose pole tips were specially designed to produce a very inhomogeneous magnetic field. The potassium atoms in this field are first aligned and are then split into two groups, one in which the magnetic moment lies parallel to the field and the other in which the magnetic moments are reversed. There is no reason why one group should be preferred to the other, so the potassium atoms are equally divided into the two groups. Now a magnetic dipole

Magnetism

experiences not only a twisting couple tending to turn it, but in an inhomogeneous field a force of translation as well, and this force will be oppositely directed for the two groups of atoms. The beam of atoms emerges from the magnetic field, therefore, split into two distinct beams in each of which the atomic magnetic moments are all pointing in the same direction. Stern and Gerlach were not only able to observe this splitting of the original beam but by allowing the atoms to fall and condense on a glass plate placed at a known distance from the electromagnet they were able to calculate the force that the magnetic field must have exerted on the atoms in order to separate the beam in two parts, and from this they were able to measure the magnetic moment of the individual atoms.

So far we have been concerned solely with atoms and their properties, but we must remember that matter as we know it consists of atoms in various states of aggregation. Very often the atoms combine to form molecules and the nature of the resulting substance depends very much on the forces between them. In gases the individual atoms or molecules are very loosely held to each other and are to be regarded as being in a state of rapid, random motion as a result of which collisions between them occur very frequently. The speed of their motion between collisions is a measure of the temperature of the gas; the higher the average speed, the higher the temperature, and collisions occur more frequently the greater its density. In air at atmospheric temperature and pressure an air molecule makes on the average 10^{10} collisions every second. Between collisions the atoms or molecules are free particles, and gases, whose magnetic properties should most closely resemble those of free atoms, would at first sight appear the ideal substances to investigate. Unfortunately gases are so weakly magnetic that experiments with them are very difficult to carry out.

If a gas is cooled the motion of the molecules becomes less and less until eventually a state is reached at which the molecules are moving so slowly that one molecule may exert a force on its neighbours for a considerable period of time. The force between the molecules causes the gas to condense and form a liquid. In a liquid the individual molecules are much more

The Atomic Theory of Matter

closely packed together than in a gas – one cubic centimetre of water at its boiling point becomes 1700 cubic centimetres of steam – but although the speed of movement of the molecules is much less than in a gas, this movement is still considerable and is still largely a random one. At still lower temperatures a liquid solidifies. In a solid the atoms or molecules are packed together in a perfectly regular manner to form a crystal, and it is worth stressing the fact that all solids are crystalline; even metals which can be cast into any pre-determined shape and apparently bear little resemblance to the crystals of quartz or copper sulphate which some of us may have seen or grown, will be found upon close examination to consist of numerous small crystals* each of which preserves the features that we normally associate with more obviously crystalline substances. A solid may crystallize in a number of different crystal structures. Some of these are very complex and their exact form can only be found by many months of tedious and painstaking measurement, but fortunately we need consider only the two simplest types. These are the so-called cubic and hexagonal crystal structures to which categories all the ferromagnetic metals belong. The atoms are packed closer together in these two structures than in any others, and it is not without significance that a large number of the elements which crystallize in these forms are metals with which a very high mechanical strength is usually associated. The two crystal structures are illustrated in Figures 13a and b. In cubic crystals the atoms are packed together to form a cubic array, and the whole crystal is to be regarded as being made up of a very large number of unit cubes packed together without any spaces between them. The size of these cubes is usually very small, their sides being about 3×10^{-8} cm. in iron, so that the finite size of the building blocks normally goes unnoticed. But by using special techniques involving very great magnification it has been possible to demonstrate the existence of 'steps' of an apparently flat sur-

* Single crystals of zinc, often quite large, can frequently be seen on the surface of galvanized iron sheet or wire. They give the surface a 'mottled' appearance due to the fact that the reflecting power of a crystal depends upon the orientation of the crystal with respect to the surface.

Magnetism

face which correspond to the size of these building blocks of atoms. There are two variants of this cubic structure because of the way it is possible to insert extra atoms without destroying the cubic nature of the packing of atoms. In the first an extra atom placed at the centre of the cube produces the so-called body-centred cubic structure (Figure 13c), which is the form in which iron crystallizes. In the second the extra atom is situated at the mid point of each cube face; this form is termed a

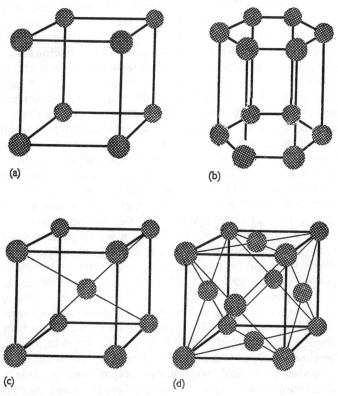

Figure 13. Some simple crystal structures. (a) Simple cubic – no known examples; (b) Hexagonal – e.g., cobalt (for simplicity, some atoms have been omitted in the drawing); (c) Body-centred cubic – e.g. iron; (d) Face-centred cubic – e.g. nickel.

The Atomic Theory of Matter

face-centred structure (Figure 13d) and is the structure possessed by crystals of nickel. In the so-called hexagonal structure the atoms are packed together to form a hexagonal array and the unit, or building block, out of which the entire crystal is formed is a hexagonal prism, as illustrated in Figure 13b.* Cobalt possesses this structure at temperatures below about 400° C, above which it undergoes a transformation to the cubic structure. The properties of cobalt change abruptly at this temperature and this emphasizes the fact that the properties of most solids, and especially metals in which the atoms are packed so closely together, are the properties not so much of the individual atoms but of the way in which they are arranged. This dependence of the properties of solids, magnetic and otherwise, on their crystal structure is due to the very strong forces of interaction between the atoms and the fact that they depend somewhat on the way the atoms are arranged. It is of course just because these forces are so strong that solids, and particularly metals, possess such great mechanical strength and so strongly resist extension and compression. The forces between atoms are primarily electrical in nature, for although magnetic forces between atoms do exist they are usually too feeble in comparison with electrical forces to exert a very great influence. As an illustration of this fact we may note that the only elements which are held together by magnetic forces alone are helium, neon, argon, etc. and the forces are so feeble that they have very low melting points and at room temperature all these substances are monatomic gases, the so-called rare gases. In metals the interactions between the atoms are so strong that not only do the atoms lose their individual identity, but the electrons lose their allegiance to the parent atom and become free to wander about through the crystal lattice. These electrons, being charged and free to move, can carry an electric current and the high electrical conductivity of metals is due to their presence. For this reason they are often known as conduction electrons. They are also partly responsible for the mechanical strength of metals so that it is no accident that metals, with their good electrical properties, are also mechanically strong.

*For simplicity some of the atoms have been omitted from this diagram.

Magnetism

Most non-metallic elements which are solids at room temperature, sulphur, phosphorus, and iodine for example, crystallize in forms other than the cubic and hexagonal structures and in which the density of packing is considerably smaller. Their properties are more closely related to the properties of their individual atoms, since the distance between them is larger than in metals. They are usually electrical insulators and mechanically weak.

CHAPTER 6

Paramagnetism and Diamagnetism

PARAMAGNETIC substances are those which are attracted towards the stronger parts of an inhomogeneous magnetic field. The force of attraction is usually quite small, although the effect is appreciable in certain strongly paramagnetic salts. For example, small crystals of manganous chloride, $MnCl_2$, will stick to the poles of a strong permanent magnet. The magnetic susceptibility of this salt is about 10^{-4}, which is about ten times larger than that of typical paramagnetic substance; and for these the order of magnitude of the force experienced in the field of a typical laboratory electromagnet would be between one hundredth and one tenth of a gram weight per cubic centimetre of paramagnetic substance. This is a very small force and special techniques are required to measure it accurately. The susceptibilities of paramagnetic salts decrease with increasing temperature, while those of metals, which are usually smaller than those of salt, are often independent of temperature. Diamagnetic substances are repelled from the stronger parts of a magnetic field and their susceptibility, apart from being negative, is usually smaller than those of a typical paramagnetic. Diamagnetic susceptibilities are usually independent of temperature. Typical diamagnetic substances are copper, mercury, and the rare gases. An exceptional example is bismuth whose susceptibility not only has an unusually large value but which varies with temperature as well.

This distinction between para- and diamagnetics was recognized by Faraday in 1845, and the interpretation of these differences in behaviour was first given by the French physicist Paul Langevin (1872–1946) exactly sixty years later. Langevin realized that the magnetic behaviour of a substance, considered as an assembly of atoms, depends first and foremost upon whether or not the individual atoms possess a magnetic moment.

Magnetism

If they do, then the substance behaves as a paramagnetic; if they do not, then purely diamagnetic behaviour results. We now know that each electron in an atom possesses a spin magnetic moment of its own and that the motion in an orbit gives rise to an orbital magnetic moment. The resultant magnetic moment of the atom is the resultant magnetic moment of all the orbital and spin moments combined. Now, it may happen, and in fact does so quite often, that the individual moments are situated in such a way as to cancel each other out and the atom as a whole has no net magnetic moment. The simplest example would be provided by an atom containing two electrons each rotating in identical orbits but in opposite directions. The magnetic moments associated with each electron would then be equal and opposite, and the total effect would be zero. It does not follow from this that the spin moments necessarily cancel each other in this way, but in fact they often do, and the atom which corresponds roughly to this picture, namely that of helium, has no magnetic moment and helium gas is in fact diamagnetic. On the other hand the individual magnetic moments may not completely annul one another, and in this case the atom is left with a resultant magnetic moment of perhaps one or two Bohr magnetons. A substance made up of an assembly of such atoms will not have any magnetic moment of its own, however, unless a magnetic field is applied, because the magnetic moments of the individual atoms are pointing in all directions, the resultant of which is zero.

Suppose a magnetic field is now applied to such an assembly. Each atom behaves like a small magnet and is rotated towards the direction of the field. The substance now becomes magnetized; it acquires a magnetic moment because the individual atomic moments align themselves parallel to the magnetic field, and like any other magnet the substance experiences a force tending to pull it towards the place where the field is strongest. This is typical paramagnetic behaviour. However, common experience teaches us that the force of attraction is quite small. If the atomic magnets were completely free to turn, the force experienced by a paramagnetic substance would be as great if not greater than that experienced by a bar of iron in similar

Paramagnetism and Diamagnetism

circumstances, and anyone who has held a screwdriver or penknife near a powerful electromagnet knows how great this can be. Evidently there is something which prevents a paramagnetic substance from acquiring such a large magnetic moment, by preventing the atomic moment from turning more than a slight amount.

Now we have already seen that the atoms in a substance at ordinary temperatures are to be regarded as being in a state of violent agitation, and that temperature can in fact be regarded as a measure of the violence of this agitation. If it were not for the continual presence of thermal agitation the atomic moments would indeed align themselves parallel to an applied magnetic field and there would be no paramagnetism; all substances would be either diamagnetic or ferromagnetic, a situation which would have serious consequences in everyday life. The precise nature of this thermal agitation depends rather upon whether the substance is a solid, liquid, or a gas, but in all cases its effect is to oppose the effect of any applied magnetic field, and since the agitation is stronger the higher the temperature, this explains in a general way why the paramagnetism of a substance usually becomes weaker as its temperature is raised.

In view of the fact that the essential nature of the thermal agitation is one of completely random motion, with the atoms flying rapidly hither and thither in all directions, colliding with each other many times a second and changing their direction of motion after each collision, it might be imagined that any hope of calculating the magnetic properties of such a system would be completely out of the question. Fortunately this is not the case, and indeed the mathematical theory of a gas consisting of an assembly of randomly moving particles had been worked out by Maxwell and by Boltzmann some forty years before Langevin's work. For it turns out that although the particles are moving randomly, in all directions and with some having velocities greater than the average and others less, all that matters is the average effect of the agitation. We observe only the average effect because the number of atoms is so enormous, and any fluctuations which arise as a result of their

Magnetism

movement are averaged out by our measuring devices.* The results obtained by Maxwell and Boltzmann were used by Langevin to calculate the magnetization of a paramagnetic gas. The restriction to gases is necessary not so much because the work of Maxwell and Boltzmann was restricted to gases, for it has a far greater generality than that, but because it is necessary first to choose the simplest situation of an assembly of non-interacting atoms, each of which is free to behave independently of the others. This situation is most closely resembled in gases and it is primarily to gases that Langevin's theory applies.

It is unfortunately not possible to give a detailed account of the theory, and we shall have to be content with a statement of the final result, which is shown graphically in Figure 14. At first the magnetization increases linearly with field, or in other words the magnetic susceptibility is independent of the field strength as found by Faraday and by Curie. In this region, according to Langevin's theory the magnetization I is related to the magnetic field H by the formula:

$$I = \frac{N\mu^2 H}{3kT}$$

in which N is the number of atoms per cubic centimetre, each of which has a magnetic moment μ, H is the strength of the magnetic field applied, T is the absolute temperature, and k is known as Boltzmann's constant. This is a fundamental constant which is a measure of the strength of the thermal agitation. Its numerical value is $1 \cdot 38 \times 10^{-16}$ ergs per degree, and any object or particle at a temperature $T°K$ possesses an inherent energy due to its temperature roughly equal to kT ergs. At room temperature T is about $300°$ and so kT is roughly 4×10^{-14} ergs. This amount of energy is small by ordinary standards, but for very small light objects it corresponds to a kinetic energy of movement involving quite high speeds, as Robert Brown

* Actually with specially sensitive apparatus these so-called thermal fluctuations can readily be observed. They were first noticed by the English botanist Robert Brown (1773–1858), who found that small pollen grains seen under a microscope appeared to be in a state of perpetual movement. Brown himself believed this motion to be due to a life force associated with living things.

Paramagnetism and Diamagnetism

observed when he noticed the movement of pollen grains under his microscope. As the field strength is increased the curve begins to flatten out just like the magnetization curve of iron, and eventually reaches a limiting value known as the saturation magnetization, which occurs when all the atomic dipoles are completely aligned parallel to the field. Increasing the field can

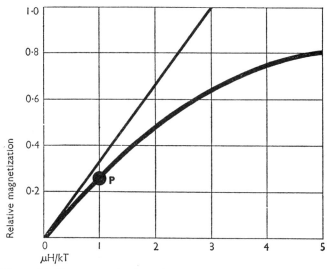

Figure 14. The magnetization of a paramagnetic gas according to Langevin's theory. The vertical scale represents the magnetization as a fraction of the saturation value corresponding to complete alignment of the magnetic moments.

have no further effect, since as we saw in the last chapter the actual magnitude of the atomic magnetic moment is unaffected by an external field. If the field is now removed the magnetization drops to zero; there is no remanence as is observed in ferromagnetic materials.*

How large a field is necessary to reach saturation in a para-

* Unfortunately for the theory remanence is actually observed in certain paramagnetic substances, albeit at temperatures lower than those so far mentioned – see page 107.

Magnetism

magnetic substance? The answer is quite simple; the field must be sufficiently large that the turning effect which it produces on an atomic dipole is greater than the disturbing effect due to thermal agitation. The turning effect on a magnetic dipole μ is proportional to μH and the energy of thermal agitation is proportional to kT, and it will be noticed that in Figure 14 the magnetization is plotted not against field alone but against the ratio $\mu H/kT$. Let us calculate the order of magnitude of the magnetic field needed to reach the point P on Figure 14. This corresponds to a value of $\mu H/kT$ equal to one. At room temperature kT is roughly 4×10^{-14} ergs and, assuming μ to be one Bohr magneton, $\mu = 9 \cdot \times 10^{-21}$ electromagnetic units. H comes out to be about five million oersteds. This is very much greater than any steady magnetic field which has yet been produced. Continuous, steady fields of up to 250,000 oersted can now be produced in special laboratories corresponding to a value of $\mu H/kT$ of about one twentieth at room temperature. This is so nearly zero that it cannot be shown on the figure. Thus at room temperature the magnetization is directly proportional to the applied field for all fields that can reasonably be obtained. At very low temperatures the situation may be quite different. At $1 \cdot 5°$ K, kT is only 2×10^{-16} ergs and so the point P could be reached in a field of about twenty thousand oersteds. Of course at this temperature all the known gases have liquified or solidified, but there are certain paramagnetic salts which exhibit typical paramagnetic behaviour at low temperatures. At high temperatures or, more correctly, for small values of $\mu H/kT$ the formula on page 86 predicts Curie's Law, for by dividing both sides by H we see that the magnetic susceptibility is inversely proportional to the absolute temperature, which is what Curie found for a large number of substances. We can now see that a substance which obeys Curie's Law is one which satisfies the assumptions made in Langevin's theory, namely that there are no interactions between the atoms, so that each atom behaves independently of the others. Thus we should expect gases, particularly the rare gases, to obey the law, as is observed. Moreover by measuring the paramagnetic behaviour of a substance we can find out whether Curie's Law is obeyed or not and hence gain informa-

Paramagnetism and Diamagnetism

tion about the existence of interactions between its atoms. Indeed by observations of the extent of its disobedience to Curie's Law we may even be able to glean information about the strength of the interactions. This is a simple example but the first we have had so far of the way in which magnetic measurement can assist in our investigations into the nature and behaviour of matter.

Before going on to discuss the magnetic behaviour of more complicated substances such as metals, it is first necessary to inquire what happens in the case of an atom whose magnetic moment is zero. At first sight it might seem that such an atom in which the orbital and spin moments exactly cancel each other out would be magnetically neutral and would not react in any way to the presence of a magnetic field. In fact this is not so, because the orbital motion of an electron is affected somewhat by the presence of a magnetic field and this causes a slight effect known as diamagnetism.

It is not difficult to see how this effect arises if we fix our attention upon a single electron revolving in a circular orbit and regard it as a single turn of wire carrying an electric current (Figure 7). When a field is applied an electromotive force is set up in the wire according to the ordinary laws of electromagnetic induction, and a current flows in such a direction as to oppose the applied field. The wire would therefore tend to be pushed away from the strongest part of the magnetic field. In a wire the current ceases as soon as the applied magnetic field ceases to change owing to the electrical resistance of the wire, which always tends to slow down the motion of the electrons inside it. But if the wire were made of a substance whose electrical resistance were zero this current would persist as long as the field remained; it would fall to zero only if the field were switched off or otherwise removed, thus setting up an equal and opposite electromotive force in the wire. So it is with the electron in an orbit. When the field is switched on the speed of rotation changes slightly and thus the magnetic moment of the orbital electron changes slightly, but always in such a way as to oppose the magnetic field that causes it to change. There is nothing like electrical resistance opposing the motion of an

Magnetism

electron in an orbit, so that this change in speed persists as long as the field is applied. There are one or two important things to notice about this effect. First, it does not matter in which sense the electron is rotating in the orbit, for in both cases the change in speed is such as to oppose the field producing it. Thus although the magnetic moments of the individual electron orbits may annul one another the diamagnetic effects add together and a small magnetic moment is always induced which is always in the opposite direction to that of the field producing it. Secondly, even if the orbital magnetic moments do not cancel each other and the atom as a whole possesses a permanent magnetic dipole moment, giving rise to paramagnetism, this diamagnetic effect still occurs as well. Usually the diamagnetism is considerably smaller than any paramagnetic effects which may be present, and this is why we ignored it in our previous discussion of paramagnetism. To demonstrate that this is so we must first estimate the magnitude of the diamagnetic effect.

For this purpose it is necessary only to calculate the change of speed of an electron in an orbit caused by the application of a magnetic field. As stated in Chapter 3 the primary effect of a changing magnetic field is to set up an electric field, which in this case can be regarded as being concentric with the electron orbit. If the electric field strength is denoted by E then the force on the electron whose charge is e is Ee and its acceleration is force divided by mass, or Ee/m. The electric field strength is equal to the e.m.f. induced in the orbit divided by the total length of the orbit; and since the e.m.f. is equal to the rate of change of magnetic flux linking the orbit, which is AH, A being the area of the orbit, which remains unchanged, we have

$$E = \frac{A}{2\pi r} \times \text{(rate of change of magnetic field)}$$

where r is the radius of the electron orbit.

Therefore the acceleration is

$$\frac{Ee}{m} = \frac{Ae}{2\pi r m} \times \text{(rate of change of magnetic field)}$$

But acceleration is just the rate of change of velocity and so the total change in velocity is

Paramagnetism and Diamagnetism

$$\frac{eA}{2\pi rm} \times \text{(total change in magnetic field)}$$

$$= \frac{eAH}{2\pi rm}$$

Now as we saw in Chapter 5 the magnetic moment M due to a charge e revolving with velocity v in a circular orbit of radius r is

$$M = \frac{eAv}{2\pi r}$$

Therefore the change in magnetic moment is equal to

$$\frac{eA}{2\pi r} \times \text{(change in velocity)}$$

$$= \frac{e^2 r^2 H}{4m}$$

If the atom contains many electrons the induced magnetic moment is just the sum of the change in the moment of each orbit. Since these orbits will in general be different in size the total induced moment is

$$M = \frac{e^2 H}{4m} \times \text{(the sum of the } r^2 \text{ for each orbit)}$$

This assumes that each orbit is at right angles to the field and so experiences the maximum effect. In practice the orbits will be orientated in all directions with respect to the field direction, and when this is taken into account the effect is to multiply the expression by a factor equal to two thirds. The susceptibility per atom is obtained by dividing the induced moment by the field and is thus

$$\frac{e^2}{6m} \times (r_1{}^2 + r_2{}^2 + r_3{}^3 \ldots)$$

in which r_1 means the radius of electron number one, r_2 that of electron number two, there being as many terms inside the bracket as there are electrons in the atom. The quantities, e, m, and r are constants independent of temperature, and so the diamagnetic susceptibility is the same at all temperatures in agreement with observation. For many atoms the radii of the various electron orbits can be calculated, and by making use of the above formula their diamagnetic susceptibilities can be obtained. Usually the calculated susceptibilities agree quite closely with those determined experimentally, especially in the

Magnetism

regularly as the atomic weights increase because of the associated increase in the number of electrons in the atom. Thus the atomic susceptibility of argon is greater than that of neon which is in turn greater than that of helium, the lightest member of the series. All other gases are composed of molecules at ordinary temperatures. When atoms combine to form molecules they usually do so in such a way as to favour the possession of completely filled electron shells. Now the possession of an even number of electrons always favours the likelihood that the spin and orbital magnetic moments will point in such directions as to annul one another, and this always happens when the electron shells are completely filled. Thus it is to be anticipated that most gases which are composed of molecules are diamagnetic, and in fact the only common gases which are paramagnetic are oxygen, O_2, and nitric oxide, NO. Oxygen is in fact quite strongly paramagnetic and closely obeys Curie's Law, and when liquified is so strongly paramagnetic that it will adhere to the poles of a strong permanent magnet.

In liquids the molecules are neither completely free as in gas nor fixed rigidly in position as in solids, and this makes it very difficult to construct a theory which will predict all their properties. The most common of all liquids, water, is diamagnetic, with a susceptibility of -0.7×10^{-6} electromagnetic units per cubic centimetre. The only elements which are liquids at room temperature are bromine and mercury, and these too are diamagnetic. The diamagnetism of bromine is independent of temperature, in agreement with theory; that of mercury, however, decreases as its temperature increases, although it is a good conductor of electricity. This anomalous behaviour is believed to be due to free electrons – which it must contain in order to account for its electrical conductivity.

The magnetic properties of solids are perhaps the most interesting of all if only because of the diversity of different types of solid materials. We are familiar with the characteristics of liquids and gases, and although we recognize differences in the properties of different gases and different liquids, to say that a substance is either a gas or a liquid is to define its properties fairly exactly. This is not the case with solids; rubber, coal, and

Paramagnetism and Diamagnetism

steel are all solids, yet apart from the fact that they each retain the shape in which they are formed their properties could hardly be more different. It is of course just this great diversity of properties which makes solids so interesting generally and not only from a magnetic point of view. Unfortunately this means that we cannot expect to formulate any hard and fast rules which will enable us to predict their magnetic properties. We shall therefore content ourselves with a few general remarks – bearing in mind that exceptions to what quasi-general rules we make can nearly always be found – and a few specific examples of substances whose magnetic properties are for one reason or another especially interesting.

Physically, the simplest type of solid is the inorganic crystal, of which sodium chloride, common rock salt, is perhaps the best known example. The sodium atom has one electron in its outermost shell, while the outermost shell of the chlorine atom needs just one electron to fill it completely. When the two atoms are brought close together, what might be termed the spare electron from the sodium atom transfers itself to the chlorine atom. The sodium atom, being now short of one electron, now loses its electrical neutrality and becomes positively charged, while the chlorine, having one too many, acquires a negative charge. Atoms that have lost or gained an electron are termed ions. The ions, being oppositely charged, attract each other according to the ordinary laws of electrostatics, and this force of attraction is sufficient to keep the sodium and chlorine ions together in solid rock salt (Figure 15). The ions thus formed and out of which the whole crystal is built are perfectly symmetrical and have electron shells which are completely filled. Such crystals, which are usually known as ionic crystals, are therefore diamagnetic. The ions of the alkali metals (lithium, sodium, potassium, rubidium, and caesium), those of the halogens (fluorine, chlorine, bromine, and iodine), and the ammonium and sulphate ions are all diamagnetic, so that all salts made up of these ions are also diamagnetic. Most ionic crystals readily dissolve in water and the resulting solution is diamagnetic as well.

However we must not imagine from this that all ions are

Magnetism

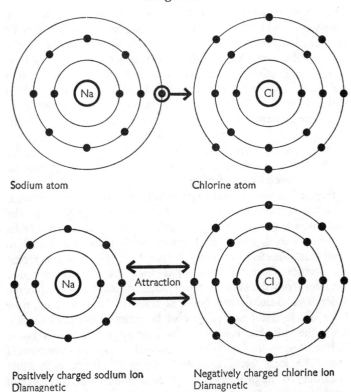

Figure 15. The formation of sodium chloride. The sodium atom loses an electron to the chlorine atom, and the resulting ions, having filled shells of electrons, are diamagnetic.

necessarily diamagnetic; indeed the simplest of all, the hydrogen ion, behaves like a paramagnetic ion. The chief paramagnetic ions are those of the elements lying between scandium and copper and including those of manganese, chromium, iron, cobalt, and nickel, and also those of that group of elements known in the periodic table as the rare earths. These elements, from cerium with atomic number 58 to ytterbium (70), have a very unusual electron configuration. The outermost shell of all the rare earths contain the same number of electrons and so

William Gilbert of Colchester, showing his experiments on electricity (electrostatics) to Queen Elizabeth I and her Court.

Hans Christian Oersted (1777–1851), who, by his discovery of the magnetic effect of an electric current, first showed the connexion between magnetism and electricity.

André Marie Ampère (1775-1836) developed the mathematical theory of electromagnetism and suggested that all magnetism might be due to circulatory electric currents.

Michael Faraday (1791-1867), the greatest experimental physicist of the nineteenth century, possibly of all time. He constructed the first electric motor and the first dynamo. With the aid of his great electromagnet (Plate 17) he discovered dia- and paramagnetism and the connexion between magnetism and light.

James Clerk Maxwell (1831–79) summarized the whole of electromagnetism in four equations and predicted the existence of electromagnetic waves.

Levitation in superconductors. Persistent currents, induced both in lead rings and in a lead sphere, all cooled in liquid helium, produce electromagnetic forces which are as great as the weight of the lead sphere, which floats 'in mid air' indefinitely.

The core of a typical 45,000 kilowatt transformer. It contains about 35 tons of silicon-iron sheet.

A 10,000 kilowatt transformer showing the core and windings in different stages of construction.

A 33,000 kilowatt high-voltage transformer.

A magnetic memory array. It contains 1,024 small ferrite rings, each of which acts as the core of a single-turn transformer.

Part of the array under high magnification showing the individual cores and their windings.

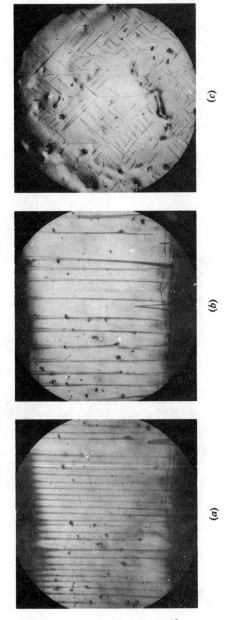

PLATES 12 AND 13: Bitter patterns showing the domains on the surface of certain ferromagnetic crystals. (a) and (b) Barium ferrite crystal. The domains are parallel strips, adjacent strips being magnetized in opposite directions. In (a) the width of the domains is almost equal and the resultant magnetization of the crystal is small. In (b) a magnetic field has been applied parallel to the domains and those magnetized parallel to the field have grown whilst those anti-parallel have shrunk almost to zero; the crystal in this condition is nearly saturated. (c) A more complicated domain structure observed in a crystal of 3 per cent silicon-iron. Here there are four easy directions, parallel to the edges of the photograph, and the direction of magnetization within each domain lies along one of these.

(a) (b) (c)

(a) A 180° wall on the surface of a 3 per cent silicon-iron crystal showing two sets of spiked closure domains around holes. (b) A 180° wall on the surface of a 3 per cent silicon-iron crystal held by two closure domains to an imperfection – probably a local stress. (c) An edge of a 3 per cent silicon-iron crystal showing three distinct types of domain. The lower left quadrant shows serrated domain walls, indicative of the existence of internal mechanical stresses. Along the upper edge triangular closure domains may be seen; the remainder of the crystal consists of domains in the form of parallel strips. The edges of the crystal are not very well defined, having been rounded off by electrolytic polishing.

(a)

(b)

PLATES 14 AND 15: Permanent magnets. (a) Silver-mounted lodestone (c.1660). (b) Permanent magnet assembly used in modern electricity meters for recording domestic power consumption. The actual magnet is shown on the left.

(a)

(b)

(a) Permanent magnet assembly used in modern ammeters. The permanent magnet, either Alnico or Alcomax, is the block at the base: the pole pieces are of rustless iron. (b) A loudspeaker magnet. The magnet is the slightly tapered cylindrical magnet shown on the right. It is magnetized radially and the magnetic flux is returned through the soft-iron annulus and outer case (cf. Fig. 39). Other forms of magnet assembly are possible but this type is very frequently used.

The great permanent magnet at Bellevue. This magnet, which is used for experiments on the magnetic deflection of α-particles, weighs 70 tons. The pole gap is variable between 2 and 10 cm. and the field from 11,000 to 6,500 oersteds.

Paramagnetism and Diamagnetism

they are chemically identical; this is why it is so difficult to obtain a pure specimen of a single rare earth element. The outside shell but two is unfilled, and this means that the resultant magnetic moment of the electrons in this shell may be quite large.

The rare earth atoms and ions are strongly paramagnetic, and salts containing their ions, and also those of the iron-group metals, are usually quite strongly paramagnetic, with a paramagnetism due to a single paramagnetic ion. Some of these salts, particularly those containing a large amount of water of crystallization, obey Curie's Law almost exactly, the reason being that the paramagnetic ions in such a crystal are often quite a long way from each other, being separated by diamagnetic ions and water of crystallization which merely serve to dilute the concentration of paramagnetic ions. The magnetic properties of such a solid resemble those of a paramagnetic gas very closely. An example often quoted is that of the salt known as potassium chrome alum. At room temperature its susceptibility is about 2×10^{-5} but at a temperature of $1°K$ this increases by about 300 according to Curie's Law to about ·007. At this temperature the substance shows paramagnetic saturation effects of the kind mentioned earlier in this chapter in connexion with an ideal paramagnetic gas. The saturation magnetization of this salt is about 60 gauss, rather more than one thirtieth of the saturation magnetization of iron at room temperature. The fact that it is as small as this, for the magnetic moment of the chromium ion responsible for the paramagnetism is more than twice that of an iron atom in metallic iron, is due to the dilution of the paramagnetic ion by the other diamagnetic ions.

Once we know that the paramagnetism of a certain salt obeys Curie's Law and can be ascribed to the presence of a single ion, it is possible to work backwards, using Langevin's formula, to calculate the magnetic moment possessed by that ion. This value should agree with that which can be calculated from our (quite detailed) knowledge of the types of electron orbits and the arrangement of the spin moments within the atom. In the case of the ions of the rare earths the agreement between the

Magnetism

calculated and experimentally determined ionic moments is very good, but with the ions of the iron group theory can only be reconciled with experiment by the assumption that the orbital magnetic moments do not contribute to the magnetization. This is not the same as saying that the orbital magnetic moment in each ion is zero; it is not. Rather is it that for some reason the orbital moments are unable to be rotated by an external magnetic field and the effective moment of the ion is due solely to the electron spins. This state of affairs in which orbital moments are completely fixed is believed to be caused by very strong interactions between neighbouring atoms. The evidence so far given for this type of behaviour is somewhat indirect and it is pertinent to ask whether these conclusions can be checked by experiment. The answer is that they can and this seems to be a suitable place to describe the relevant experiment, not only because it ranks among the classical experiments in magnetism, but because we shall need to refer to the results when we come to deal with ferromagnetism.

The experiment was first suggested by Einstein and carried out in collaboration with de Haas. It is based upon the knowledge that in an atom the electrons are rotating in orbits and spinning on their axes, so that each electron possesses a certain amount of angular momentum. In atoms of paramagnetic substances in which the orbital and spin moments do not annul each other the angular momentum is not annulled either, and each atom can be regarded as a kind of spinning top with angular momentum and an associated magnetic moment. In an unmagnetized paramagnetic substance the atomic moments of all the atoms are arranged at random so that their net effect is zero; likewise the total angular momentum of the substance is zero. When a magnetic field is applied the atomic moments tend to align themselves along the field direction. But this means also that the axes of the atomic spinning tops become slightly aligned and their total angular momentum no longer averages out to zero. Now there is a fundamental law in physics which states that angular momentum cannot just appear but is always conserved in any physical process. If the substance originally possesses zero angular momentum when unmagnetized its

Paramagnetism and Diamagnetism

total angular momentum must remain zero when magnetized, and in order to bring this about the substance as a whole acquires an angular momentum equal and opposite to that of the slightly aligned atomic magnets inside it. The substance therefore tends to twist about an axis parallel to the applied magnetic field and will do so if freely suspended. The experiment consists therefore simply in suspending a substance from a fine fibre and measuring the amount of twist when it is magnetized. The word simply in the previous sentence requires some qualification because the experiment, although simple in principle, is extremely difficult to carry out because the effect is so minute.

The magnitude of the effect depends chiefly on the magnetization which the substance can acquire, and since this is greater for ferromagnetic than for paramagnetic substances Einstein and de Haas suggested that the experiment be first tried with iron. This they did, and on magnetizing a long rod of iron were able to observe the twisting effect which they anticipated. The theory of the effect predicts that for a given change in magnetization the twisting effect if the magnetism is due to orbital motion of the electrons should be twice that if magnetism is due to electron spin. Subsequent highly refined experiments by Chattock and Bates, working at Bristol, showed conclusively that the magnitude of the twisting effect in iron is such as to indicate that the whole of the magnetization is due to electron spin, and this has also been found to be the case for nickel and for cobalt.

The difficulties in experiments of this kind are immense, and it might be expected that it would be out of the question to perform them with paramagnetic substances in which the maximum magnetization which they can acquire at room temperature and with the field strengths commonly available is only about one hundred-thousandth of the saturation magnetization of iron. However, by a special refinement of the technique Sucksmith has succeeded in doing this. Sucksmith's measurements show that in the paramagnetic salts containing ions of the iron group of elements the magnetism is due entirely to electron spins, and the orbital moments, though not zero, are

Magnetism

completely unable to orientate themselves in an external magnetic field. In the rare earth ions on the other hand the magnetization is due to a combination of both orbital and spin moment, and by measuring the magnitude of the twisting effect in salts containing these ions it is possible to find out how much of the magnetism is due to spin and how much due to orbital effects.

These so-called gyromagnetic effects may be observed in another way – backwards as it were – by rotating an unmagnetized substance and observing the magnetization it acquires by rotation alone, and experiments of this kind have been successfully carried out by Barnett in the U.S.A. Once again the experimental difficulties are very considerable, as may be appreciated from the fact that a speed of one hundred revolutions per second produces the same magnetization as would a magnetic field of one ten millionth of an oersted. No additional information is given by these experiments, since they are just the converse effect of that predicted by Einstein, but they provide a useful check on the more direct methods.

We leave paramagnetic crystals for a while (we shall come up against them again in Chapter 12) and make a few brief remarks about solids which do not conduct electricity, in other words insulators. Most of the elements which are insulators are feebly diamagnetic, typical examples being boron, silicon, phosphorus, and sulphur. Usually these solids crystallize in rather open structures, so that the distance between atoms is quite large and interaction effects are usually unimportant, so that many of them are magnetically similar to an ideal diamagnetic gas. A large number of oxides and sulphides are diamagnetic, but as might be expected oxides and sulphides of the iron group of elements and of the rare earths are anomalous. Oxides of the rare earths are usually strongly paramagnetic, while the oxides of the iron group have even more complex properties.

The mineral pyrrhotite, whose chemical formula approximates to FeS, is also ferromagnetic, as are also a number of compounds of manganese with arsenic and bismuth. We shall have occasion to refer to these compounds in Chapter 7, and

Paramagnetism and Diamagnetism

their appearance here serves only to remind us how difficult it is to make general rules about the magnetic properties of solids.

Organic compounds, those composed of carbon, hydrogen, and oxygen, are almost invariably diamagnetic. In a comprehensive series of experiments the French physicist P. Pascal found that the molecular susceptibility of many organic compounds is the sum of the susceptibilities of the atoms of which they are composed plus additional contributions which are characteristic of the type of chemical bond which holds the atoms together within the molecule. The rules formulated by Pascal can often be used in reverse to find out the nature of chemical bonds in molecules whose magnetic susceptibilities have been measured. The information obtained in this way is particularly useful to the chemist when for one reason or another normal chemical methods of investigation have failed to provide it.

The diamagnetic susceptibilities of many complex organic molecules, particularly the aromatic hydrocarbons formed from benzene, is unusually high. Benzene itself is an example, and the solids naphthalene and anthracene are others. It will be recalled that the benzene molecule is composed of six carbon and six hydrogen atoms arranged in a hexagonal ring as shown in Figure 16 (which also shows the structures of naphthalene and anthracene) and it has been suggested that this is due to electrons which become detached from their parent atoms and, instead of pursuing their normal circular orbits within the atom, wander completely round the benzene ring. Since the contribution of an electron to the diamagnetic susceptibility is proportional to the square of the radius of its orbit it is clear that the value of r^2 for a complete benzene ring is much greater than for a normal circular orbit, and consequently the diamagnetism of such molecules is very large. It is clear too that the magnitude of the diamagnetic effect will depend very much on the orientation of the benzene ring with respect to the applied magnetic field, being a maximum when the face of the ring is perpendicular to the field and zero when parallel. Thus by measuring the diamagnetic susceptibility of a crystal of anthracene in various directions one can gain valuable

Magnetism

information about the arrangement of the benzene rings within the crystal.

The most complex solids of all are the metals, and the reader will be prepared for the statement that metallic substances may

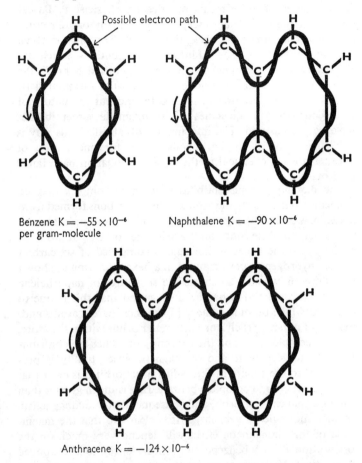

Benzene $K = -55 \times 10^{-6}$ per gram-molecule

Naphthalene $K = -90 \times 10^{-6}$

Anthracene $K = -124 \times 10^{-6}$

Figure 16. The structure of benzene, naphthalene, and anthracene. If an electron is sufficiently free to wander completely round the ring, its value of r, in the formula on page 91, is unusually large and the diamagnetism correspondingly great.

Paramagnetism and Diamagnetism

exhibit diamagnetism, paramagnetism, or ferromagnetism. In metals the atoms are packed very closely together, and the very strong interaction between them causes some of the outermost electrons to be freed from the atom which then becomes a positively charged ion. These electrons are more or less free to move throughout the metal crystal and often behave rather like a gas of free electrons; indeed by applying mathematical formulae originally derived for gases to this so-called electron gas one can often predict properties such as electrical and thermal conductivity which are in close agreement with those actually observed.

In the rare earth metals the ions are themselves paramagnetic and so the metal as a whole is paramagnetic, usually very strongly. The magnetic moment of the rare earth ions is so large that they interact quite strongly and in the presence of a magnetic field the field acting upon an ion is composed of the external field plus an additional field due to all the other ions as well. If we make the reasonable assumption that this field is proportional to the net magnetization of the substance I, then the total field acting upon an ion is not just H but $H + \alpha I$, in which α is a constant. If we use this expression for the total field in Langevin's formula for paramagnetism we obtain

$$I = \frac{N\mu^2}{3kT}(H + \alpha I)$$

which, if we solve for I, becomes

$$I = \frac{N\mu^2 H}{3k(T - N\mu^2\alpha/3k)}$$

The susceptibility I/H is thus of the form

$$\kappa = \frac{C}{T - \theta}$$

where C and θ are constant. This is just the Curie-Weiss Law mentioned on page 57 and which is closely obeyed by the rare earth metals. The internal field due to the ions represented by the term αI always helps to increase the paramagnetism and when T becomes equal to θ the susceptibility becomes infinite and we have magnetization without any external field. Such a

Magnetism

situation corresponds to ferromagnetism, and many rare earth elements do in fact become ferromagnetic at low temperatures. However this subject properly belongs to the next chapter, and we shall defer any further discussion of these elements and those of the iron group until then.

In metals the magnetic properties arise from three different origins. Firstly there is a part due to the metal ion. This, more often than not, is diamagnetic. For example when sodium atoms form the metal, the one electron which causes chemical combination and makes sodium such a reactive element becomes the itinerant electron. It is responsible for carrying electricity and is known as a conduction electron. In sodium just one electron per atom is free to become detached, and consequently there are as many conduction electrons in metallic sodium as there are sodium atoms. The remaining sodium ion possessing completely filled electron shells is diamagnetic and this is also true of most metal ions (platinum is an exception to this rule). Secondly there is the magnetism due to the conduction electrons. Since there are in any metal approximately as many conduction electrons as there are atoms it might at first sight appear that these should give rise to strong paramagnetism, since each carries a spin moment of one Bohr magneton. However, metals are not usually any more strongly paramagnetic than insulators, and this was for a long time very puzzling. The difficulty was resolved just over thirty years ago by Fermi and Dirac, who first showed that the electrons, on account of their unusually small mass, obey certain special laws which prevent more than a minute fraction of them from being affected by a magnetic field. The number of electrons which can be affected turns out to be virtually independent of temperature, and so the paramagnetism of the conduction electrons, in addition to being small, differs from that due to orientation of atomic magnets and which obeys Curie's Law, or some modification thereof, by being almost unaffected by temperature. Finally there is one other effect due to the conduction electrons. In a metal these are all moving about, randomly, and with quite large velocities. A magnetic field causes them to travel between collisions in curved paths which are arcs

Paramagnetism and Diamagnetism

of a circle similar to the way in which a magnetic field causes a charged particle to move in a circular path in a cyclotron (see page 228). But an electron travelling in a circular path is equivalent to a current flowing in circular wire and has a magnetic moment. The induced moment produced by the bending of the electron's path is by Lenz's law in such a direction as to oppose the action of the field, and so the effect produced is diamagnetic. If the electrons are perfectly free, theory shows that this diamagnetic effect should be just one third of the paramagnetic effect and, like it, almost independent of temperature.

The total magnetic properties of a metal thus depend on the relative magnitudes of the three contributions. Sometimes the combined diamagnetic effects prevail and the metal as a whole is feebly diamagnetic as in copper and zinc. In others the paramagnetic effect is paramount as in the alkali metals, e.g. sodium.

More complex metals, especially those known in the periodic table as transition metals, are paramagnetic owing to the large magnetic moments of the metal ions. The magnetic susceptibilities of all the elements in the periodic table are summarized in Figure 17.

From a magnetic standpoint the exception among the metals has always been bismuth, which has so far been omitted from the discussion on that account. It is a metal with a very strong diamagnetism (volume susceptibility -13×10^{-6}). Bismuth loses its unusually high diamagnetism when it melts, and this suggests that the large diamagnetism is due to the structure of the metal rather than any peculiarity of the bismuth atom itself. The theory of the magnetic properties of bismuth has been worked out mathematically and it turns out that the large diamagnetism is due to a peculiar combination of the effects of the crystal structure and the number of electrons in the bismuth atom. Thus it is something of an accidental combination of circumstances that gives bismuth its unusual properties. Tellurium, which has chemical properties similar to those of bismuth and which possesses the same crystal structure, does not possess the right number of electrons to exhibit the effect at all strongly and is only feebly diamagnetic.

This brief survey of the magnetic properties of solid materials

Magnetism

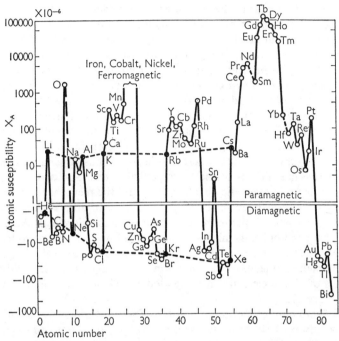

Figure 17. The magnetic susceptibilities of the elements.

would hardly be complete without some reference to the very remarkable properties exhibited by certain substances at very low temperatures. One effect, paramagnetic saturation, has already been mentioned on page 87. In itself there is nothing exceptional about this phenomenon, which merely shows that for an ideal paramagnetic substance which obeys Curie's Law, what requires enormous magnetic fields at room temperature can be achieved in fields of a few thousand oersteds at very low temperatures. It is noteworthy that at low temperatures a number of these paramagnetic salts which show saturation effects also exhibit characteristic ferromagnetic properties such as remanence and possess a coercive force different from zero. Figure 18 shows a hysteresis loop obtained at a low temperature with the salt, iron ammonium alum.

The exact reasons for ferromagnetic behaviour in para-

Paramagnetism and Diamagnetism

magnetics are not very well understood at present, but it must be due to some sort of cooperative action between the atomic magnetic moments which results in their being coupled together in such a way to prevent them from springing back to their original position when the field is removed.

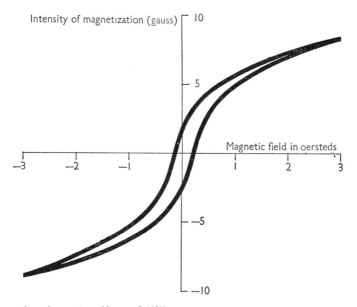

Iron Ammonium Alum at 0·03°K

Figure 18. Hysteresis loop in iron ammonium alum at 0.03°K.

At low temperatures, usually below 10° K, a number of metals lose their electrical resistance and become perfect conductors of electricity – they are termed superconductors. The loss of electrical resistance occurs quite abruptly at a critical temperature, which depends on the nature of the metal; for example, lead becomes superconducting at 7·26° K.

Some of the consequences of a substance's having no electrical resistance are quite remarkable. For example, a current in a superconductor does not require an electromotive force to maintain it as is the case with ordinary conductors. Conse-

Magnetism

quently a current, once started in a superconductor, persists indefinitely. By way of illustration, suppose we have a ring of lead maintained at a temperature slightly greater than the critical temperature, so that it is still an ordinary conductor. A bar magnet pushed towards the ring induces an electromotive force in the ring, and a current flows which by Ohm's Law ceases immediately the e.m.f. ceases, i.e. as soon as the magnet comes to rest. This is because of the electrical resistance of the metal lead, which causes the movement of the electrons to be rapidly slowed down owing to a kind of friction which the electrons experience in moving through the metal, and in which, like all forms of friction, the original energy of movement of the electrons is ultimately dissipated as heat in the metal.

If the same experiment is now done with the lead ring cooled below its critical temperature there is no resistance to slow down the motion of the electrons and the induced current (which is actually very closely confined to the surface) persists until the magnet is removed and an e.m.f. is induced which, being equal and opposite to the first one, just nullifies it. Currents induced in superconducting rings have been found to persist for many hours without appreciable diminution in strength.

It will be observed that the induction of currents in a superconducting ring is exactly the same in principle as that which leads to diamagnetism due to an electron in a circular orbit, and as might be expected superconductors are very strongly diamagnetic. If instead of being a ring the superconductor is solid the induced currents are more properly termed eddy-currents, but the effect is still the same, namely to give rise to a strong diamagnetic effect. This effect is so strong that the magnetic induction inside a superconductor is actually zero, the induced currents being sufficiently great that their effect neutralizes that of the magnetic field. Thus $H + 4\pi I = O$ or $I = -H/4\pi$. The diamagnetic susceptibility I/H is therefore equal to $1/4\pi$, roughly one twelfth, which is about ten thousand times the susceptibility of bismuth. If a superconducting hollow sphere is placed in a non-uniform magnetic field it will experience a force tending to push it away from the strongest parts of the field. By suitable arrangement of the magnetic field it is possible to make this force act upwards so that levitation

Paramagnetism and Diamagnetism

occurs and the sphere remains suspended above the field 'without any visible means of support' (see Plate 6). Actually although we have explained the large diamagnetism of superconductors in terms of their lack of electrical resistance it seems to be an independent property. For if a superconductor is cooled below its critical temperature in the presence of a constant magnetic field it is found that the induction is still zero inside the metal in the superconducting state. Somehow the magnetic lines of force must get pushed out at the same time that the substance loses its resistance, but this cannot be caused by the effects of electromagnetic induction, because the magnetic field is kept constant during the transition from the normal to superconducting states and therefore by Faraday's Law no electromotive force is induced. It is only in fairly weak magnetic fields that superconductors show diamagnetism, and once a certain critical field strength is exceeded superconductivity is destroyed, leaving the normal dia- or paramagnetism of the metal.

Almost immediately after the discovery of superconductivity the possibility of using a coil of superconducting wire to produce large magnetic fields was suggested. However, it was soon pointed out that each turn of wire of the coil would be subjected to a magnetic field approximately equal to that produced by the whole coil, and if this happened to be greater than the critical field it would destroy the superconductivity. The critical field for a typical superconductor is a few hundred oersteds, and this is the maximum field that a superconducting coil can generate if it is to remain superconducting. Recently, however, a group of superconducting alloys having very large critical fields has been discovered. One of these, an alloy of niobium and tin, has a critical field between 200,000 and 300,000 oersteds at $0°K$, and superconducting magnets cooled in liquid helium can now be constructed capable of producing continuous magnetic fields up to 150,000 oersteds. Once energized these magnets require no further expenditure of energy and are, despite the large consumption of liquid helium required to cool them, particularly useful when large magnetic fields are required over large volumes and for long periods of time.

CHAPTER 7

Ferromagnetism

THE word ferromagnetism is used to describe the properties and behaviour usually associated with the metal iron. We have mentioned some of these properties in Chapter 4, and have already noted that the elements nickel and cobalt are also ferromagnetic. Since only a few of the hundred or so elements now known are ferromagnetic our first task will be to inquire into the special combination of circumstances which gives rise to ferromagnetism.

The first explanation of ferromagnetism was given in 1907 by Pierre Weiss, who put forward the hypothesis that ferromagnetism might be the result of an unusually strong interaction between the individual atomic magnets, which in some way made them all point in the same direction. Weiss supposed that the field acting upon an iron atom is composed of the external field plus an internal field due to the resultant effect of the magnetic moments of all the other atoms. If these moments are all pointing in different directions their effect at any point will cancel out and the field at that point is just the external field. But if they are rotated somewhat and become slightly aligned along the direction of the field this is no longer the case. Weiss assumed that the field at any point is now equal to the external field plus an internal field. Since this internal field is zero when the substance is unmagnetized and cannot get any larger when the atomic magnets are completely aligned it is taken as being proportional to the magnetization of the substance. The total field is thus $H + \alpha I$, where α is a constant. This is the way in which interactions between the magnetic moments of paramagnetic substances are formally taken into account, and as we have seen this leads immediately to the law of Curie-Weiss for the magnetic susceptibility, viz.

Ferromagnetism

$$\kappa = \frac{N\mu^2}{3k(T - N\mu^2 a/3k)}$$

$$= \frac{c}{T - \theta}$$

As long as T is greater than $N\mu^2 a/3k$ this susceptibility varies inversely, not as the temperature, but as the difference between the temperature T and some other temperature, say θ, equal to $N\mu^2 a/3k$. Purely paramagnetic behaviour has already been discussed in Chapter 4 and there is no need to pursue this further. But what happens when T becomes less than θ?

A moment's consideration will show that the appearance of the term θ in the denominator is a direct consequence of the presence of an internal field due to interactions, for if there are no interactions, $\alpha = 0$, $\theta = 0$, and the equation reverts to Curie's Law. The effect of the interaction is to decrease the denominator, with the result that for a given field the magnetization is increased. The two terms θ and T in the denominator represent respectively the tendency towards alignment of the atomic magnets in an orderly manner by the internal field, and the tendency towards random orientation of the atomic magnets by the disordering effect of thermal agitation. For temperatures greater than θ, the effect of thermal agitation is paramount and the substance is paramagnetic, usually quite strongly, owing to the interaction which tends to increase the magnetization. Below θ however the internal field is decisive in determining the orientation of the atomic magnets which become aligned parallel to each other by the internal field. This is not to say that thermal agitation is ineffective at temperatures below θ, but that its role is subordinate to that of the internal field in determining the magnetization of the substance.

Gyromagnetic experiments indicate quite clearly that in ferromagnetics the magnetism is due almost entirely to electron spin, and we shall henceforth use the word spin rather loosely, to mean the elementary carriers of magnetism in ferromagnetics. We can visualize the situation inside a ferromagnetic substance somewhat as follows. At temperatures greater than θ the spins, although not without influence upon each other, are too violently agitated by the effects of temperature to behave cooperatively. Each spin behaves more or less independently and the substance is paramagnetic. As the temperature is lowered the motion of the atoms becomes feebler and the spins are given a little more time, as it were, to take notice of each other. Eventually, when the temperature θ is reached, the effects of temperature have become so small in comparison with the effects of the forces of interaction that two neighbouring spins become parallel to each other for appreciable periods of time. As soon as this happens the internal field due to the two parallel spins is increased (the magnetic moment of two parallel spins is twice that of a single one) and this increased internal field acts on those spins in the immediate vicinity causing them to become aligned parallel to the other two. The process, once started, proceeds very rapidly as a result of the cooperative action between spins, and the alignment process spreads rather like an advancing avalanche throughout the whole substance. This alignment of the spins takes place solely on account of the internal field and without the presence of any external field, and consequently the magnetization acquired by a substance as a result of this process is usually referred to as its spontaneous magnetization. The temperature θ at which spontaneous magnetization sets in is known as the Curie temperature.

The tendency towards perfect ordering of the spins is always opposed by thermal agitation, and consequently the alignment is not absolutely perfect except at the absolute zero of temperature. Nevertheless, so strong are the internal forces that in iron the spontaneous magnetization at 680° C, only 90° below its Curie point, is already slightly more than half its value at absolute zero. A curve showing how the spontaneous magnetization of iron and nickel vary with temperature is shown in Figure 11. Curves for all other ferromagnetic substances are very similar.

What determines the magnitude of the spontaneous magnetization in a substance? Obviously it depends upon the temperature, although, as Figure 11 will show, the spontaneous magnetization of a ferromagnetic substance is rather insensitive to temperature except near its Curie temperature. It is

Ferromagnetism

virtually unaffected by an external magnetic field, however strong, a fact which we shall appreciate later. Since the magnetism is due to electron spin only, the only other factor which can determine the magnetization is the number of electron spins per cubic centimetre of the substance which are capable of being aligned. In iron there are on the average 2·2 electron spins per atom each possessing a magnetic moment of one Bohr magneton, 0.93×10^{-20} electromagnetic units. The spontaneous intensity of magnetization is thus $2.2 \times 0.93 \times 10^{-20} \times$ the number of iron atoms per cubic centimetre. Now a cubic centimetre of iron contains about 8.3×10^{20} atoms and so, multiplying these three numbers together we get a value of 1,700. In nickel, in which there are only an average 0·6 electron spins per atom, the spontaneous magnetization at low temperatures comes out to be about 500. The highest spontaneous magnetization, about 1,900 gauss, is possessed by the alloys of iron and cobalt containing between 35 per cent and 40 per cent cobalt (but see p. 118).

This account of spontaneous magnetization has so far tacitly ignored any discussion of the nature of the interaction forces which bring it about. The Curie temperature θ is equal to $N\mu^2 a/3k$ and, as we should expect, is itself a measure of the strength of the interaction expressed by means of the constant a. We have spoken as though the interactions were purely magnetic in origin, similar to those which would exist between an array of freely suspended compass needles. If this were the case it is quite easy to show mathematically that $a = 4\pi/3$, or roughly 4. If we substitute this value in the above expression for θ and use appropriate values for N and μ we obtain a temperature of about one tenth of a degree absolute for the Curie temperature of iron. Clearly, magnetic interactions cannot by themselves account for the observed Curie temperature, which is about ten thousand times as great as this. In fact it is one of the most remarkable features of magnetism in general that the interaction effects, which have played such an important part in our discussion of paramagnetism and which by a simple extension of the theory account for the broad features of ferromagnetism in so satisfactory a manner, are not due to

Magnetism

internal magnetic forces at all but are primarily electrical in nature. Weiss, in his original formulation of the theory of ferromagnetism, was obliged to postulate a value for α of the order of 100,000 in order to account for the observed Curie temperature of ferromagnetic substances. Such a value of α corresponds to an internal magnetic field of over 10^7 oersteds* so we can appreciate that the spontaneous magnetism is unlikely to be much affected by the magnetic fields which we can produce in the laboratory which only rarely exceed 50,000 oersteds.

The large value of α remained something of a mystery for just over twenty years until an explanation was found by the German physicist W. Heisenberg, using the methods of quantum mechanics. Heisenberg's theory is too involved to reproduce here, but we can perhaps give a very brief account of its underlying principles. If we consider a system containing two electrons only, such as the hydrogen molecule, we should expect the forces between its constituent nuclei and electrons to be due solely to the electrostatic forces of attraction between the negatively-charged electrons and positively-charged nuclei and repulsion between the two electrons. According to the quantum theory this is still true, but it is not the whole story, for one of the features of the quantum theory is its insistence upon the indistinguishability of electrons. It is no use labelling the electrons in the hydrogen molecule, since there are no means of telling one from another, even in principle. When this fact is incorporated into the mathematical equations governing the motion of the system the effect is to add another term to the final expression for the energy of the form $-J(S_1 \times S_2)$. This energy is called the exchange energy from the fact that the electrons can change their identities without our being aware of it, and in the above expression J is a constant usually termed the exchange integral (for reasons which we cannot explain here) and S_1 and S_2 are the spin magnetic moments of the two electrons. If we recall that magnetic moment is a vector quantity we shall appreciate that the product $S_1 \times S_2$ will be positive

* There is no contradiction here: the internal forces are *not* due to magnetic effects, but there is no harm in calculating what order of magnitude the internal field would have to be if they were.

Ferromagnetism

if the spins are parallel and negative if they are antiparallel. One of the basic rules which govern all physical systems is that a system always exists whenever possible in the state in which its energy is least, and so whether the spin moments are parallel or antiparallel depends upon which arrangement is the one of lowest energy, and this is determined entirely by the sign of the exchange integral J. In the case of the hydrogen molecule Heisenberg showed that J is negative and so the exchange energy is negative and thereby reduces the total energy of the system only if the spins are antiparallel. The spin moment of molecular hydrogen is thus zero and hydrogen gas is actually diamagnetic, as previously mentioned.

Now the sign of J depends amongst other things on the distance between atoms, and Heisenberg was able to show that in iron, nickel, and cobalt in metallic form the distances were such as to make J positive, in which case the interaction due to this exchange mechanism favours parallel spins. Further refinements of the theory which cannot be discussed here show that for ferromagnetism to occur it is not sufficient that J be positive; it must exceed a certain critical value, otherwise the substance will be merely strongly paramagnetic.

Heisenberg's theory, though not altogether free from objection (the step from the hydrogen molecule to an array of several million iron atoms arranged in a regular lattice is too large to be taken in a single stride) does explain a number of observations rather neatly. Immediately preceding iron in the periodic table is manganese, which is quite strongly paramagnetic. This is because the manganese atoms are situated at such a distance from each other that J although positive is not large enough to promote ferromagnetism. If nitrogen is introduced into the manganese lattice this has the effect of pushing the manganese atoms further apart and increasing the numerical value of J. Four nitrogen atoms to every hundred are all that is necessary to make the resulting substance quite strongly ferromagnetic with a Curie temperature of about 500° C. Similarly quite a large number of compounds of manganese are ferromagnetic, notably those with bismuth, arsenic, phosphorus, sulphur, and tin. The metal alloys known as the Heusler alloys containing

Magnetism

manganese, aluminium, and copper, but no ferromagnetic element, are likewise ferromagnetic, as are similar alloys in which the copper is replaced by silver. Chromium, which comes just before manganese in the periodic table, behaves similarly, and alloys of chromium and tellurium and also with platinum are also ferromagnetic.

The Curie temperature of most ferromagnetic metals changes with pressure. We should expect this to occur, since by applying pressure the distance between the atoms is decreased slightly and so the value of J will alter. This is equivalent to changing the value of α on Weiss's theory and so θ changes. In most metals the change of Curie temperature with pressure is quite slight, but in an alloy of nickel and iron containing 30 per cent of nickel the effect is so great that at room temperature the alloy ceases to be ferromagnetic at all.

The occurrence of ferromagnetism in solid materials is, as a result of recent research, much less rare than was at one time believed, and new examples are constantly being added to the list. All alloys of the ferromagnetic elements with each other are ferromagnetic as also are alloys of the magnetic elements with non-magnetic metals, albeit over a restricted range of compositions. Some of these are of technical importance whose properties will be discussed later. For example iron-silicon and iron-aluminium alloys containing up to 32 per cent and 19 per cent by weight respectively are ferromagnetic as are nickel copper alloys containing 0–60 per cent of copper. Usually the effect of adding a non-magnetic metal to iron or nickel is to decrease both the spontaneous magnetization and the Curie temperature, so that a number of these alloys may appear non-magnetic merely because their Curie temperature is below room temperature. The ferromagnetism of the rare earth elements has already been mentioned and their magnetic properties are summarized in the table on page 117. In addition certain cyanides (salts of hydrocyanide, or prussic acid) containing iron, cobalt, and nickel ions, amongst which are the well-known dyes Prussian blue and Turnbull's blue, are also ferromagnetic at temperatures below about 40° K. One of the most recent discoveries of this kind has been the ferromagnetism of certain alloys of

Ferromagnetism

zinc and zirconium. While it is unlikely that alloys such as these will ever be of any practical or commercial interest there is little doubt that their discovery and the investigation of their properties will lead to a better understanding of how and why substances are ferromagnetic.

The Ferromagnetic Elements

Element–Atomic Number in brackets	Curie Temperature °K	Saturation Intensity of Magnetization at 0°K
Iron (26)	1043	1740
Cobalt (27)	1393	1430
Nickel (28)	631	510
Gadolinium (64)	292	2110
Terbium (65)	221	2710
Dysprosium (66)	85	3000
Holmium (67)	20	3080
Erbium (68)	20	2420
Thulium (69)	22	2190

The forces which are responsible for aligning the spins in a ferromagnetic substance below its Curie temperature are so strong that it is hardly surprising that the occurrence of spontaneous magnetization should be accompanied by other physical changes as well. Electrical resistivity and mechanical properties both show quite sudden changes at the Curie temperature. Of greater importance however are the changes in dimensions which are associated with the onset of spontaneous magnetization. For example, nickel contracts on cooling through the Curie temperature. The total expansion or contraction due to the onset of the ferromagnetic state is proportional to the square of the spontaneous magnetization, so that the changes take place most rapidly just below the Curie temperature where the spontaneous magnetization is varying most rapidly. The extra expansion or contraction is of course superimposed on the normal thermal expansion, and so the total thermal expansion of a ferromagnetic substance becomes either abnormally great or abnormally small near the Curie temperature according as the spontaneous magnetization is accompanied

Magnetism

by a contraction or an expansion (Figure 19). The unusually low thermal expansion of Invar (36 per cent nickel, 64 per cent iron, Curie temperature 26° C) is an example of this effect put to practical use. Needless to say the low expansion of Invar is restricted to a narrow temperature range but by adding other metals it is possible to increase this range. One such alloy, Super Invar, expands by only about 25 parts in a million on being heated from 0° C to 100° C. This is only slightly greater than the expansion of brass for 1° C. Such alloys find numerous applications, for example in pendulum clocks where the cumbersome temperature compensated solid pendulum can be effectively replaced by an Invar rod, and in internal combustion engines where distortion due to expansion at elevated temperatures of operation must be avoided. A

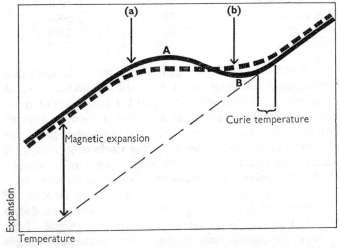

Figure 19. A ferromagnetic substance may contract or expand when spontaneous magnetization occurs below its Curie temperature. If it expands then there may be two small temperature ranges over which the total thermal expansion is very small, as at A and B (curve a). By adding other elements an alloy can be produced in which these two temperature ranges coalesce into one larger one (curve b). Note that on curve (a) the expansion is negative between A and B.

Ferromagnetism

further important class of applications of low-expansion alloys is for purposes of temperature control. The rod type of thermostat used to control the temperature of the domestic cooking oven is an example. The principle of operation here is the relative movement between a rod having a low coefficient of expansion such as Invar, and those of a normal material such as can be used to actuate a control valve which will cut off the gas supply when the required temperature is reached. Other types of thermostat make use of two metals of different thermal expansion coefficients, welded together along their entire length, thus forming a strip somewhat like a long single-decker sandwich. Such a strip may be straight at room temperature but will become curved on heating, and if one end is fixed the movement of the free end may be used to operate electrical switches. This type of thermostat is frequently used when space is limited and is most commonly found in the controlled electric iron. The final application which should be mentioned is that of glass-to-metal seals for the leads of electric lamps and radio valves. These have to be brought out through the glass envelope, and it is very important that the envelope be properly sealed to prevent entry of air into the evacuated enclosure and that the glass is not strained by differential expansion between itself and the metal leads, thereby causing brittleness and risk of fracture. This can be achieved by finding an alloy whose thermal expansion is the same as that of the glass used, and alloys of iron and nickel containing between 46 per cent and 50 per cent of the latter satisfy these conditions very well. It is interesting to note that until the development of these alloys for this particular purpose the only metal which could safely be sealed into glass was platinum, and large savings have been made possible through the replacement of platinum by these nickel-iron alloys.

The impression must not be given that everything is now known about the occurrence of ferromagnetism in metal and alloys. Some attempt to dispel such a view has been made by quoting examples of alloys, such as those of zinc and zirconium, whose ferromagnetism is hardly to be expected. In fact although the salient features of ferromagnetism are understood

Magnetism

the details are still lacking. When, earlier, the statement was made that in iron each atom contributes an average 2·2 to the magnetism the obvious question is: Why 2·2 electrons per atom of iron and only 0·6 per atom of nickel?*

The fact is we do not really know the answer to this question. On account of the enormous complexity of a solid substance it is impossible to investigate it mathematically. True, the correct equations could be formulated, but no one could expect to solve them in a single lifetime. So for the time being we have to be content with mathematic approximations which often amount to little more than plausible guesses, testing the validity of our guesses by comparison with experiment. Until such time that the great computers can solve all our equations for us this is all we can do, hoping ultimately to reach the correct answer by a slow process of trial and error. Meanwhile there remains a vast store of facts yet to be discovered and coordinated into a satisfactory theory.

* In fact the figures quoted are derived from the experimentally determined values of the spontaneous magnetization at low temperatures. Recent experiments employing the diffraction of neutrons confirm them within the limits of experimental error.

CHAPTER 8

Ferromagnetic Domains

THE theory of Weiss and its later development by Heisenberg is remarkably successful in explaining the way in which spontaneous magnetization occurs in ferromagnetics. It is, however, incomplete in one important respect, for it leads to the conclusion that every ferromagnetic substance below its Curie temperature should, on account of the spontaneous magnetization, be a permanent magnet. Actually it is quite possible to have a piece of iron in an unmagnetized condition, a fact which is known from common experience. In order to explain this Weiss, with remarkable foresight, introduced the so-called 'domain hypothesis'. He supposed that a ferromagnetic material such as iron is divided into domains, each of which is spontaneously magnetized to a degree appropriate to its temperature. Within each domain the magnetization is constant both in magnitude and direction, and to this extent a domain does resemble a permanent magnet somewhat. In order to account for the fact that the net magnetization of a piece of iron is zero it is necessary to assume that the iron contains a large number of these domains and that their magnetic moments point in different directions, so that their effects annul each other when there is no external magnetic field acting upon them. The domains, although small compared with ordinary sizes, are nevertheless large enough to contain very many atoms; in fact by atomic standards the domains are very large indeed and consequently are almost unaffected by the effects of thermal agitation. Thus ferromagnetic substances are much easier to magnetize than paramagnetics; the effect of a magnetic field is to rotate the magnetic moment of the domains (not the domains themselves, which, being composed of atoms, are rigidly fixed in the crystal lattice of the substance) until they all lie parallel to the field. The substance is then said to be saturated (see page 59), and we can now appreciate that the saturation magnetization of a

Magnetism

ferromagnetic substance is identical with the spontaneous magnetization within each domain. Some ferromagnetic substances saturate much more easily than others, and one of the questions we shall try to answer in this chapter is why it is that Permalloy (an alloy of iron and nickel) becomes almost completely saturated in a field of less than one tenth of an oersted, while a substance like cobalt is not fully saturated even in a field of ten thousand oersteds. In order to do this it is necessary to know a little more about these domains, their size and shape and the factors that govern their arrangement.

The first indications of the correctness of the domain hypothesis came in a very indirect manner from an experiment due to a German engineer, H. Barkhausen, in 1919. He found that the magnetization curves of ferromagnetic substances are not quite smooth but proceed in a jerky irregular manner. The jerks are usually quite small, and in order to observe them it is necessary to examine the magnetization curve under high magnification. Barkhausen did this by winding a coil of many turns of wire round an iron rod and connecting this coil through an amplifier to a pair of earphones. When a gradually increasing field was applied to the iron rod, thereby causing its magnetization to change slowly, a rustling noise was heard in the earphones. When the field was increased very slowly indeed this rustling sound was found to be made up of a succession of sharp clicks. These clicks are caused by the generation of a momentary electromotive force in the coil on the iron rod. If the magnetization increased smoothly there would be no clicks but merely a small steady e.m.f. induced in the coil by the changing magnetic flux linked with it. The clicks must be due to changes of magnetization taking place suddenly within the iron, and Barkhausen interpreted this as being due to the sudden swinging round of the direction of the magnetization within a domain. We now know that this interpretation is not entirely correct, but it provides a rough picture of what is happening inside a ferromagnetic substance which is being magnetized. Barkhausen's experiment was one of the first to take advantage of the triode valve, which had been discovered and patented a few years earlier. This provides a simple example

Ferromagnetic Domains

of the close relation between technological developments and scientific discoveries, for without the triode valve, which formed the basis of his amplifiers, the experiment of Barkhausen could not have been carried out and Weiss's hypothesis would have remained a mere hypothesis for much longer.

One can hardly regard the Barkhausen effect as direct confirmation of the existence of domains, and it was not until ten years later that the American physicist, F. Bitter, suggested a method whereby ferromagnetic domains might be made visible. The idea behind this method is that finely divided ferromagnetic particles, sprinkled over the surface of a ferromagnetic substance, might reveal the presence of domains much in the way that iron filings sprinkled around a magnet arrange themselves so as to reveal the magnetic lines of force. For his ferromagnetic particles Bitter originally used finely divided iron suspended in a liquid of low viscosity so that they settled quickly, but the more recent practice is to use a colloidal solution of magnetite, a magnetic oxide of iron otherwise known as lodestone. A drop of colloidal magnetite is placed on the surface of the material under investigation, which is then examined with a microscope. It is found that the colloidal particles collect along lines separating two domains and thus conveniently mark out the domains on the surface. It must be admitted that the early results obtained using this technique were not too successful. The domain patterns can be seen only on a perfectly flat polished surface, preferably on a single crystal of a ferromagnetic metal. If the surface is polished mechanically it is left in a strained state owing to the scratches made by even the finest abrasive, and the patterns obtained are characteristic of a strained surface layer rather than the domains of a strain-free crystal. Fortunately the depth of this strained layer is quite small, of the order of a few thousandths of an inch, and this can be easily removed by a method known as electrolytic polishing which is very similar to electro-plating in reverse. When this is done the true nature of the domain patterns is revealed by the magnetic colloid and valuable information concerning the sizes and arrangement of ferromagnetic domains can be obtained in this manner.

Magnetism

The really successful application of this technique to the study of ferromagnetic domains may be said to have begun about 1947 and is due chiefly to the work of L. F. Bates and his collaborators at the University of Nottingham, and of H. J. Williams in the Bell Telephone Laboratories in the United States. These studies have contributed enormously to our knowledge of domains and of the manner in which a ferromagnetic substance becomes magnetized. Some examples of the beautiful patterns which may be observed are shown in Plates 12 and 13.

Most of our knowledge of domain shapes and sizes has come from the application of the magnetic colloid technique to the study of single ferromagnetic crystals. As the examples show, the domains are usually regular structures often possessing quite a simple shape. In order to understand why this should be so it is necessary to mention some of the magnetic properties of single crystals. The use of single crystals of ferromagnetic substances as a means of investigating the magnetic properties of more common polycrystalline substances is important, as it illustrates an important aspect of the method of scientific inquiry which is to draw general conclusions from the behaviour of simple systems, and to apply these conclusions to the behaviour of more complicated systems. For as mentioned on page 79, metals are usually polycrystalline, being composed of many thousands of interlocking crystals all pointing in arbitrary directions. The properties of such an aggregate are therefore the average properties of all these crystals (often called crystallites or crystal grains) in the different directions. In order to obtain any real understanding of the behaviour of ordinary materials the simplest and best way is first to investigate the behaviour of single crystals. Techniques for growing large crystals of metals have been developed since the Second World War and a large number of investigations on single crystals have been carried out.

When one comes to examine the magnetic behaviour of single crystals one finds that the properties vary somewhat with the direction of the crystal in which these properties are measured. Figure 20 is an example and shows the magnetization curve of a single crystal of iron measured with the magnetizing

Ferromagnetic Domains

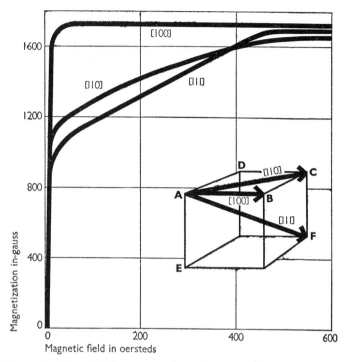

Figure 20. Magnetization curves along the three principal directions of a single crystal of iron.

field along a cube edge (called by crystallographers the [100] direction), along a face diagonal, [110], and along a body diagonal, [111]. The three magnetization curves are quite different and it is noticeably easier to magnetize an iron crystal along a cube edge than along a body diagonal. These two directions are often known as the directions of easy and difficult magnetization respectively. In order to account for this behaviour there must be present in the crystal strong forces which make it much easier for the spontaneous magnetization to point along a cube edge than a body diagonal. These forces must of course exist in addition to the strong exchange forces which align the electron spins and give rise to spontaneous

Magnetism

magnetization. A substance whose characteristics depend upon the direction in which they are measured is said to be anisotropic* and the internal forces which make one of the crystal directions a preferred one are usually known as anisotropy forces. We can measure the magnitude of the anisotropy forces by measuring the field strength necessary to produce saturation in either the [110] or [111] direction, that is the field required to turn the spontaneous magnetization from a [100] direction into one of these two, in which case the necessary field strength is directly proportional to the magnitude of the anisotropy. A more direct method is to apply a large magnetic field to a single crystal in an arbitrary direction sufficiently great to saturate it. The crystal will tend to twist in such a sense as to try and make the nearest easy direction coincide with the direction of the applied field. The twisting force is proportional to the strength of the internal anisotropy force and this can be determined by measuring its strength. The magnetization of nickel crystals is also anisotropic (as are all ferromagnetic crystals), but in this case the easy directions are the body diagonals and the difficult ones are the cube edges, just the opposite of the behaviour of iron.

Single crystals of cobalt possess a hexagonal structure as shown in the insert of Figure 21, and in this case the form of the magnetic anisotropy is rather simple. The long, hexagonal axis is the only direction of easy magnetization and instead of a difficult direction there is a difficult plane at right angles to this. It is interesting to note that the anisotropy of cobalt changes with temperature in such a way that above 300° C the hexagonal axes become the difficult direction and the basal plane becomes an easy plane. This fact is of some importance, for it indicates that the anisotropy must be due to a complex combination of interactions which is so delicately balanced that its effect can be reversed merely by altering the temperature. At present we have only a rough idea of what causes magnetic anisotropy but the behaviour of cobalt is enough to indicate that it cannot be due to a single straightforward interaction such as, for example, the exchange interaction which is respon-

* From the Greek *an* – without, *iso* – equal, *tropos* – turning, inclination.

Ferromagnetic Domains

sible for ferromagnetism. Whatever causes the anisotropy forces they are very much weaker than the exchange forces, for the latter is equivalent to an internal field of about 10^7 oersteds whereas the equivalent field which secures the spontaneous magnetization to the easy axes is rarely more than a few thousand oersteds.

One final remark needs to be made, and this is that the nature of the anisotropy of ferromagnetic crystals depends upon the way the atoms are arranged in the crystals. The hexagonal axis

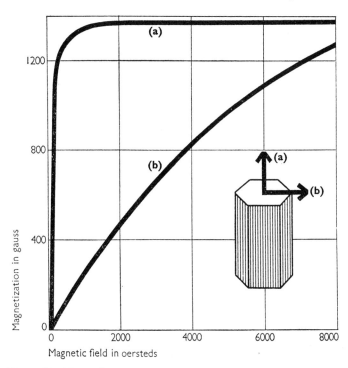

Figure 21. Magnetization curves of a single crystal of cobalt, (a) parallel and (b) perpendicular to the hexagonal axis. Notice that a very much larger field is necessary to saturate cobalt along a 'difficult' direction than iron. This fact is expressed by the statement that the anisotropy of cobalt is greater than that of iron.

Magnetism

in cobalt is a unique axis from a crystallographic viewpoint, which must be a direction of difficult or easy magnetization and not intermediate between them. In a similar manner the easy directions in iron and nickel are along some simple direction within the crystal passing through a line of atoms and thus closely related to the crystal structure of the material. In all cases the type of anisotropy exhibited reflects the way in which the atoms are arranged in the crystal lattice.

In order to illustrate the formation of domains let us first consider the very simple case of a block of some hypothetical ferromagnetic substance, initially magnetized to saturation by the application of a strong magnetic field. What happens when the field is removed and why? This question was first posed (and answered) by the Russian physicists, L. Landau and E. Lifshitz, in 1935 in a classic paper which pioneered the mathematical study of domains in ferromagnetic crystals. If the substance is saturated it means that all the atomic magnetic moments are aligned, the block consists of one huge domain, and the substance is a permanent magnet. Like any other permanent magnet, the poles at each end give rise to a magnetic field which as demonstrated on page 53 is a demagnetizing field tending to reduce the magnetization of the block. An equivalent statement is that since the block is situated in its own magnetic field, which is oppositely directed to that of its magnetization, the potential energy of the system is high. If the magnetic block were situated in an external magnetic field of strength H the energy would be equal to the work done in rotating it through one quarter of a revolution from the direction in which its magnetic moment is parallel to H. This can be shown to be $I \times H$ per cubic centimetre. If however the field is provided by the magnet itself the energy is $\frac{1}{2}$ I.H. Furthermore the demagnetizing field of a permanent magnet is equal to DI where D is a constant known as the demagnetizing factor which depends only on the shape of the body. The energy of a magnet in its own demagnetizing field is therefore $\frac{1}{2}DI^2$.

Now we have already seen that a system will always go from a state in which its potential energy is high to one in which it is low, if it is possible for it to do so. Is it therefore possible to

Ferromagnetic Domains

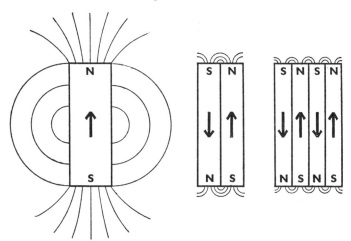

Figure 22. Illustration of the reduction of the demagnetizing field and energy by the formation of domains. Each successive subdivision confines the lines of force more closely to the substance and thereby reduces its energy. In this and subsequent figures the arrows represent the direction of magnetization within each domain.

reduce the energy of the ferromagnetic block in a simple manner? Obviously it is, by dividing it up into two halves each magnetized to saturation but in opposite directions, as shown in Figure 22. It is easy to see that the demagnetizing field is reduced by this means and we should be prepared for the statement, which can be verified mathematically, that the division into two domains, since they are both saturated, has reduced the energy by one half. Further subdivision is evidently possible, and the mathematical theory shows that if the block is divided up into N layers alternately magnetized in opposite directions the original energy is reduced by N. Why therefore does the process not continue indefinitely so that each layer is just one atom thick and the energy is reduced to its smallest possible value?

The answer to this question is to be found in the nature of the transition layer between two adjacent domains. The direction of magnetization from one domain to the next does not change

Magnetism

abruptly, but gradually, and a certain amount of energy has to be expended in the formation of this layer, which is known as a domain wall or domain boundary. The energy required to form it may be regarded as energy stored in the wall, and so although the division into domains reduces the magnetic energy every new domain wall increases the total energy of the domain walls, and there comes a stage at which the reduction in magnetic energy by domain formation is equal to the increase of energy necessary in order to enable another domain wall to form. No further subdivision is possible because this would result in an increase in the energy of the system, which cannot happen unless we supply this from an external agency.

Now this is not the only way in which the domains may be formed, for as Landau and Lifshitz showed, it is possible to

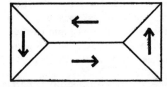 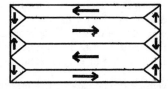

Figure 23. Some possible domain structure with closure domains. The formation of closure domains at the ends closes the magnetic path completely and prevents the formation of any external magnetic field. The state of affairs may be likened to a permanent magnet with 'keepers'.

envisage a slightly different mode of domain formation in which the demagnetizing field is zero. This can be achieved by the formation of triangular domains at the ends which carry the magnetic flux from one of the main domains to the other in a closed path within the block of the material. Figure 23 shows two possible arrangements. The small domains at the end are known as closure domains since they serve to close the magnetic flux within the material, thereby avoiding the demagnetizing field arising from the magnetic poles which would otherwise appear at the end.

We must now say a little more about the domain walls that

Ferromagnetic Domains

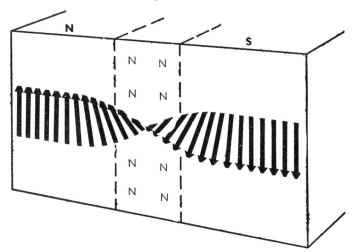

Figure 24. The nature of a ferromagnetic domain boundary. The magnetic moments (spins) rotate gradually from one domain to another in the direction shown.

form the transition layer between domains. The nature of this layer was first studied mathematically in 1932 by the German physicist, F. Bloch.

To be specific let us consider part of a ferromagnetic crystal containing two neighbouring domains in which the spontaneous magnetization is oppositely directed. Within each domain the spins are parallel and lie along the easy axes of the crystal. Between the domains the spins change direction gradually, as shown in Figure 24. The change of direction is gradual and not abrupt because the exchange forces favour parallel spins everywhere and much less energy is involved in a slow transition than in an abrupt one. The exchange forces thus try to make the transition layer as wide as possible. On the other hand the anisotropy forces try to retain the magnetic moments along one or other of the easy directions and thus favour a layer as narrow as possible. The thickness of the layer is thus determined by the competing claims of the exchange and anisotropy forces. A large anisotropy such as is observed in

Magnetism

cobalt ensures that the domain boundary is quite narrow – about one thousand atoms thick – while with a low anisotropy the width may be greater by a factor of ten or a hundred. Conversely it turns out that the energy of formation of the boundary is greater for substances with large anisotropy than for those in which it is small. For iron the energy is about one erg per square centimetre of boundary surface. Consequently we can appreciate that more domains can be formed more easily when the anisotropy is low than when it is large.

We are now also in a position to understand how the magnetic colloid technique works, for if we picture a domain wall intersecting the surface of the ferromagnetic crystal we see that within the domain itself the magnetization is everywhere along the surface. There are no magnetic poles and hence no magnetic field at the surface other than that which may be applied externally. Within the domain wall however the spins change direction, and at the centre of the wall the magnetization is actually directed at right angles to the surface. This is equivalent to a line of magnetic poles which give rise to a strong magnetic field at the surface which magnetizes the colloidal particles and attracts them towards the region where the field is most intense, namely at the centre of the domain boundary

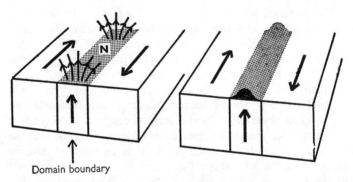

Figure 25. The formation of domain patterns. At the centre of each domain wall the magnetization is at right angles to the surface. This gives rise to a magnetic field which attracts the magnetic colloid to its strongest part, thereby delineating the domain boundary.

Ferromagnetic Domains

(see Figure 25). The colloidal particles collect therefore along lines where the domain boundaries intersect the surface and conveniently map out the domains for us.

So far we avoided any reference to a specific ferromagnetic substance or to a definite crystal, and we must next find out what determines the sizes of the domains and their arrangement in any actual case. Before doing this it is well to remember that domains form to reduce the energy associated with the demagnetizing field, and since this is determined largely by the shape of the substance, being small for a long thin rod and becoming larger the shorter and fatter the rod is made, it is evident that the domain structure will depend upon the shape of the substance. Consider for example a single crystal in the form of a long thin bar with an easy axis along the length. The poles developed at the ends are so far apart that the demagnetizing field is negligible and so the bar remains a single domain. As the bar is shortened the poles become closer together and the demagnetizing field increases. Eventually the time comes when it is advantageous to form domains, and we then get one of the formations shown in Figure 23. We cannot say which will be more likely at this stage without going into the subject in considerably greater detail, but it depends upon the demagnetizing field again, that is upon the shape of the bar. If there is no easy direction of magnetization along the bar the situation becomes much more complicated, but once again it is found that the size of the domains depends upon the size of the crystal, actually as the square root of the distance between its sides. Thus it may be seen that the domain shapes and sizes are not fundamental characteristics of the substance like its spontaneous magnetization, but depend in a complicated manner upon the shape of the crystal and the way in which the crystal axes are situated within it.

Our next task is to find out how these domains behave in the presence of a magnetic field, in other words how they determine the shape of the magnetization curve. Owing to the fact that the domain structure depends upon the shape of the crystal it is often possible to make crystals in which the domain structure is very simple. The best known example is that of a

Magnetism

rectangular 'picture frame' cut from a single crystal of iron so that each of the four straight sections is along a direction of easy magnetization. The way in which this is done is shown in Figure 26. Such a crystal, being closed magnetically, has no demagnetizing field and so there is no tendency to form domains. When such a crystal is cooled from above its Curie temperature it is found to be saturated and to consist of four domains which carry the magnetization completely round the

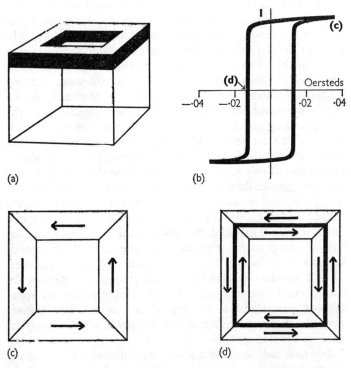

Figure 26. A 'picture frame' cut from a single crystal of iron (a), showing its domain structure (c) when saturated and (d) when the magnetization is zero. Each limb of the picture frame is along a cube edge, which in iron is an easy direction. The magnetization curve (b) arises from the movement of a single domain boundary.

Ferromagnetic Domains

crystal. If a coil is wound on the crystal and a magnetic field is applied in a direction opposing its magnetization, each domain becomes split into two by the formation of a 180° wall* which runs right round the crystal. As the field is increased the wall moves so as to increase the size of that domain whose magnetization is parallel to the field, at the expense of the other domain whose magnetization is oppositely directed. The movement of this wall can be observed by means of the magnetic colloid technique and its position can be correlated with the measured magnetization of the whole crystal. This extremely simple experiment carried out in 1947 in the U.S.A. by Williams and Shockley emphasizes two important facts; firstly that the domain walls are mobile, and secondly that a large part of the change of magnetization takes place by their movement. Similar experiments with crystals in which the domain structure is less simple have confirmed that the effect of a magnetic field is always to induce movement of the domain boundaries, which means that those domains whose spontaneous magnetization makes only a small angle with the magnetic field expand, while those in which the angle is large diminish in volume and eventually shrink to zero.

We can now interpret the magnetization curves of the single crystal of iron shown in Figure 20. Let us assume that they were measured on a crystal in which no particular shape had been chosen to ensure a simple domain structure. In the demagnetized state we can imagine the crystal to consist of a large number of domains, in each of which the magnetization lies along one of the easy directions. There are six of these in iron, and since we are assuming that all are equally favoured we can take it that one sixth of all the domains have their spontaneous magnetization pointing along one of these, one sixth along another, and so on. If we apply a field along one of these easy directions the domain boundaries move in such a way that

---

*Domain walls are classified according to the angle between the magnetization of the domains on either side. A 180° wall separates two domains in which the directions of magnetization differ by 180°. Other types of wall are possible. e.g., in Figure 23 (left) there are one 180° wall and four 90° walls.

Magnetism

the domains parallel to the direction of the field expand while all the others shrink to nothing. The magnetization increases rapidly to saturation in very small fields, as is observed experimentally. Now suppose that the field instead of being applied along an easy direction is directed along a [110], along AC for example. This time the same kind of boundary movement will occur until only two types of domain are left, those whose magnetization point along the two easy directions AB and AD. Each of these makes an angle of 45° with the field direction and the magnetization is the spontaneous magnetization multiplied by cos 45° which is $1/\sqrt{2}$ or roughly 0·7. To produce any further increase of magnetization quite large fields have to be applied, and then the spontaneous magnetization in each of the two sets of domains rotates out of the easy directions towards the field direction. Eventually they become completely parallel to the field and saturation is achieved once again, but the field required to achieve this is quite large. Turning the magnetization out of the easy directions into the field direction requires the expenditure of work against the internal anisotropy forces and this work has to be supplied by the magnetic field. For iron a field of about 500 oersteds is necessary to produce saturation in this direction. In a similar manner, if the field is applied along a [111] direction such as AF, boundary movement occurs in very small fields until there are just three sets of domains magnetized along AB, AD, and AE. The magnetization is then $1/\sqrt{3}$* of its saturation value and can likewise be increased only by the application of a magnetic field sufficiently strong to rotate the spontaneous magnetization within each domain into perfect parallelism with the field direction. A similar kind of analysis holds good for the magnetization curves of nickel crystals. Cobalt crystals have only two easy directions of magnetization. In these, magnetization along an easy direction is caused by boundary movement between domains magnetized parallel and antiparallel to the field, while magnetization at right angles to this direction is difficult from the start, the process being one of magnetization rotation from the smallest fields

* Each [111] makes an angle of 54° with a cube edge and $1/\sqrt{3}$ is cos 54°.

Ferromagnetic Domains

right up to saturation. The curves are shown in Figure 21. We may sum up these magnetization curves in the following manner. As long as there are domain boundaries to move large changes in magnetization take place in quite small fields, but as soon as it becomes necessary to rotate the direction of the magnetization out of the easy directions the magnetization increases much more slowly with the field.

Usually the ferromagnetic materials that we have to deal with are polycrystalline, being composed of a very large number of single crystal grains. The observed magnetization curve is familiar (Figure 10 for example), and we can appreciate from our discussion of single crystals that the initial steep rise of magnetization is due to growth and shrinkage of domains by movement of domain boundaries, that the knee of the curve corresponds to the state in which the favourably oriented domains in each crystallite have reached their maximum size, and that the subsequent, much more gradual increase of magnetization is due to rotation of the spontaneous magnetization in these domains, out of the easy direction into the field direction. The domain hypothesis thus explains the general form of the magnetization curve in a satisfactory manner. However in this simple form it fails to predict one of the most characteristic features of ferromagnetics, namely hysteresis, and it also fails to explain why in weak magnetic fields some materials are significantly easier to magnetize than others. Indeed our discussion of single crystals with its emphasis on the exact equivalence of the easy directions within a crystal would lead just to the opposite conclusion, that boundary movements would occur freely in low fields, permitting the magnetization to change from one easy direction into another without hindrance of any kind. The magnetization curves of all substances both single crystal and polycrystalline should all be equally steep* in low fields, which is contrary to observation. Evidently we have to look for factors which will impede the movement of domain boundaries. These are not hard to find, and investigation shows that there are two which, more than any others, conspire to prevent the wall from moving freely. These are the

* Strictly, infinitely steep.

Magnetism

effects of impurities and of local stresses and strains within the substance, and in a sense they can both be regarded as being due to some kind of imperfection in the crystal lattice which destroys, over a localized region, the regular packing of the atoms. We shall have more to say about this later, but for the moment we shall have to digress somewhat in order to describe one remarkable phenomenon exhibited by ferromagnetic substances which is of great importance in determining their behaviour in small fields. This is the phenomenon known as magnetostriction.

When a ferromagnetic substance is magnetized it is found to change length slightly. This change in length varies in magnitude from one substance to another and according to the nature of the substance it may be either an expansion or a contraction. A graph of change in length plotted against magnetic field strength shows saturation effects similar to those exhibited by magnetization curves and a given substance has a definite saturation magnetostriction just as it has a definite saturation magnetization. We speak of substances which contract on being magnetized as having negative magnetostriction and vice versa, although this terminology is of limited use in substances like iron which contract in weak fields and expand in strong ones. Magnetostriction has several important applications, some of which we shall encounter later, but for the present purpose it is not the effect itself which is important but its implications. For the external effects of magnetostriction are precisely similar to those obtained by mechanical tension and compression, and if magnetization causes mechanical effects, then conversely mechanical pressure or tension applied to a ferromagnetic substance should affect its magnetization. Quite reasonably we should anticipate that if a substance extends when magnetized, then stretching it by a tensile force should cause the magnetization to increase and vice versa. Thus the effects of stress depend upon whether the substance has a positive or a negative magnetostriction. Figure 27 shows the type of behaviour which is observed experimentally, and it agrees qualitatively with our predictions. It is noticeable that when the stress becomes very large, the material, although poly-

Ferromagnetic Domains

crystalline, behaves like a single crystal of the cobalt type. Tension applied to a substance with positive magnetostriction produces easy directions along the direction of tension, while compression makes the directions at right angles into easy directions as in cobalt above 300° C. Similar remarks apply *mutatis mutandis* to substances in which the magnetostriction is negative.

The relevance of this digression to our original problem of magnetization in weak fields becomes apparent when we

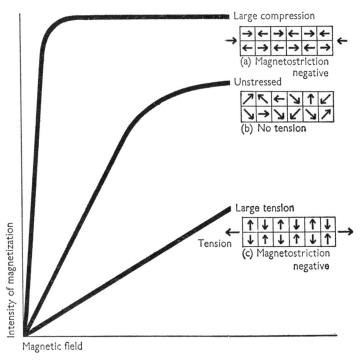

Figure 27. Magnetization curves of nickel, which contracts when magnetized (negative magnetostriction) under tension and compression. Note the similarity between the upper and lower curves and those for cobalt crystals, Figure 21.

Magnetism

realize that any actual crystal is always imperfect to a certain extent. Even in the purest crystals there are always present foreign atoms of impurities, which, being different in size from the proper atoms, do not fit properly into the crystal lattice and strain it slightly. In addition to the presence of foreign atoms there may be some dislocation of the lattice due to an atom having become dislodged from its correct position and having to squeeze in where there is no proper place for it. These displaced atoms are known to occur in all solids and their effect is to decrease the distance between the atoms in some places and to increase it slightly in others. Thus we can regard the crystal as being compressed in some places and stretched in others. The existence of these local compressions and dilations means that there are no longer six equivalent easy directions in a crystal of iron. Some are made more easy than others by these internal stresses and strains. Just how much more easy they become depends upon the magnitude of the internal stresses and the way they are distributed, and also on the magnitude of the magnetostriction, for we can appreciate that the effect of the local stresses is exactly similar to that described on page 138 in connexion with bulk material except that it is on a much smaller scale.

In the absence of an external magnetic field the positions taken up by the domain walls are determined partly by the magnetic anisotropy, but also by the extra easiness of certain directions caused by local internal stresses. When a field is applied the wall moves, but only slightly because it now has a definite preference for its original position. We can picture the wall as being held in place by a spring which exerts a restraining effect and prevents it from moving freely when acted on by a magnetic field. For small magnetic fields the displacement of the wall is elastic, proportional to the magnetic field, and so the magnetization is directly proportional to the field applied. The ratio I/H for small fields is called the initial susceptibility, and a substance with a low initial susceptibility is therefore one in which the walls are strongly held in place by what we may term the local anisotropy due to the internal stresses, and vice versa. Conversely if we want a material to have a very large initial

Ferromagnetic Domains

susceptibility we must eliminate the local anisotropy as far as possible. First of all the internal stresses must be relieved as much as possible by annealing, that is by heating the substance to a high temperature at which the atoms can rearrange themselves more easily, and cooling it very slowly to room temperature. By this means we can eliminate all the man-made stresses due to manufacture and mechanical handling. Unfortunately this treatment, though necessary, is not sufficient, for it does not get rid of stresses around foreign atoms, and it turns out that we can never completely eliminate the presence of misplaced atoms by annealing. More than this, when the substance is cooled through its Curie temperature it becomes spontaneously magnetized and magnetostriction occurs within the individual domains. This causes further straining of the crystal with the result that even after the most careful annealing process there exist certain internal stresses which cannot be eliminated completely, and which set an upper limit to the initial susceptibility. The only way of increasing this upper limit is to try and arrange matters so that these internal stresses, although present, do not have any effect. This could be done if we could make the magnetostriction zero, for then tension would be without effect on magnetic behaviour. Unfortunately we cannot get rid of the magnetostriction of a substance any more than we can destroy its spontaneous magnetization, but it is possible to find certain alloys in which the magnetostriction is very small, and such alloys do in fact develop very high initial susceptibilities when carefully annealed.

The initial susceptibilities of these alloys though very large is by no means infinite, and one may reasonably ask whether there are other factors which limit the steepness of the initial part of the magnetization curve. It turns out that there are, for the impurities which are inevitably present, besides producing internal stresses, also inhibit the ease of domain wall movement in their own right. We know this to be true from a variety of experimental evidence, but we are at present not too certain how they impede wall movement as a good deal depends on the way in which the impurity atoms are arranged in the lattice. Often they become grouped together to form a large cluster and

Magnetism

then the problem is relatively straightforward, since the atoms most likely to be present as impurities will be the non-magnetic atoms of carbon, silicon and magnesium,* and a cluster of these atoms will just act like a hole in the metals. Now if a hole is cut inside a magnetized body, magnetic poles are formed at the surface and these give rise to internal magnetic fields and demagnetizing energy in just the same way that the magnetic poles on the end surfaces of a magnet give rise to demagnetizing energy (see page 128). In that case the energy of the system is

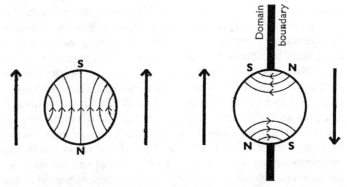

Figure 28. The reduction in internal demagnetizing energy of a small spherical hole by being intersected by a domain boundary. As a result the domain boundary becomes locked in place by its allegiance to the hole.

reduced by the formation of domains, and a similar reduction in energy takes place inside a ferromagnetic material, if a domain wall is situated so as to bisect the hole as shown in Figure 28. A wall in such a position is therefore particularly stable and resists movement which would remove it from the hole. If the hole is very large internal closure domains may form around it as in Figure 29, and this, too, can provide a means of anchoring the domain wall in position.

In both cases theory indicates that the extent to which the walls are held by the impurities is proportional to the anisotropy

* These substances, together with certain of their oxides, are very likely to be picked up during the melting stage of manufacture.

Ferromagnetic Domains

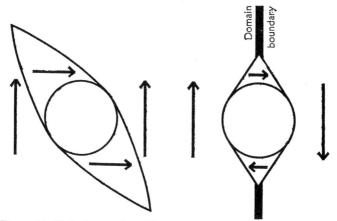

Figure 29. If the internal cavities are large enough closure domains form round them and serve to anchor the domain boundary in position.

of the substance, so that in order to obtain the highest initial susceptibilities we have to find a substance in which both the magnetostriction and the anisotropy are both as small as possible. By very good fortune it so happens that such a substance exists. It is an alloy of iron and nickel containing about 78 per cent nickel, and after careful purification it attains an initial susceptibility of about 10,000, which is about one hundred times greater than that of ordinary iron.

Everything is now accounted for (at least in principle, for the detailed analysis of the effects of impurities and internal stresses by mathematical means is fraught with the usual troubles associated with the enormous complexity of the situation) except hysteresis. Can this be adequately explained in terms of the movement of domain walls? In principle it can, although theory cannot yet provide a mathematical formula for a complete hysteresis loop. It can however indicate what the factors are which cause hysteresis to be present.

We first note that the hysteresis loop becomes closed above the knee of the curve. This means that rotation of the spontaneous magnetization out of the easy directions is reversible –

as soon as the magnetic field is removed the magnetization springs back into its easy direction. Consequently we need only look to domain wall movement for the source of hysteresis.

In our discussion of initial susceptibility it was suggested that a domain wall is held in position by forces which can be likened to elastic springs. We all know what happens when an elastic spring is stretched too much – it snaps and exerts no further restoring force on the object to which it was attached. Something similar to this occurs when a magnetic field which is sufficiently large to cause appreciable movement of the domain wall is applied. When the domain boundary is tethered by non-magnetic inclusions or holes and the field is increased the domain boundary is moved further away from the centre of the inclusion. If it is held in a few places only, the wall may become bowed slightly, but as the field is increased the time will come when it is right at the edge of the inclusions. It needs only a slight displacement, occasioned by a slight increase in field, to move it away altogether, and when this occurs the wall breaks free and moves until it is stopped by another set of inclusions. The sudden freedom of the domain wall and its subsequent rapid unimpeded motion produces a sudden increase in the magnetization, and this, it will be remembered, is just what Barkhausen found from his experiment.

If the field is now removed there will be no tendency for the wall to return to its original position, since the forces that stopped it from moving further in the presence of a magnetic field effectively retain it in that position when the field is removed. Thus we have the phenomenon of remanence. In order to make the wall go back to its original position it is necessary to apply a magnetic field in the reverse direction, and if this is increased a state will be reached when the wall breaks away from the hold imposed upon it by the inclusion and springs back to its original position once more. The field which has to be applied to bring this about is the coercive force, and thus we see that the coercive force is roughly equal to the field necessary to cause a domain wall to break away from its position of equilibrium and to move irreversibly to another similar position. We should expect that the factors which favour a

Ferromagnetic Domains

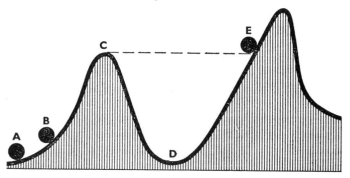

Figure 30. The 'ball and hill' analogy of hysteresis. For explanation see text.

large initial susceptibility are those which tend to make the coercive force small, and vice versa, and indeed experiments confirm that deliberate addition of foreign impurity atoms or of strains into the substance by mechanical working decreases the initial susceptibility and at the same time increases the coercive force.

We can picture the mechanism of hysteresis by reference to the analogy of a ball being pushed up a rough hill, illustrated in Figure 30. The position of the ball can be taken to be a measure of the position of the domain boundary, and this will initially be at some point such as A. If we now apply a steady force (a steady wind for example) to the ball it will roll up hill to some point B. On removal of the force the ball will naturally return to its original position. If we increase the force the ball will move further uphill until it reaches the top and then a very slight increase in the force will cause it to run down the other side of the hill and up the next one, since the force on the ball is being maintained, to the point E. This part of the ball's journey corresponds to the discontinuous change of magnetization observed by Barkhausen. The size of the sudden jump in magnetization depends upon how far the wall, or rather the ball, travels before meeting a hill which it cannot surmount. If it should happen that the height of all subsequent bumps is less than that of the first one then the ball will move a long way

Magnetism

without any increase in force. In magnetic terms this means that a large change in magnetization will take place without any further increase in magnetic field. Under these circumstances the sides of the hysteresis loops become very steep and the loop itself becomes almost perfectly rectangular. We now know how to bring this about in quite a large number of alloys, and magnetic materials possessing rectangular hysteresis loops have a number of important applications. More usually the ball will be stopped by the next bump or the next but one, and a further increase of force is necessary to move it over the top and down the other side. Reverting once again to magnetic terminology, this means that the domain boundary thus moves in a series of jerks until it meets another boundary coming towards it from another expanding domain, when the two annihilate each other and the whole crystal is magnetized in a single easy direction.

This ball analogy provides us with an insight into one other feature of hysteresis, namely the fact already mentioned on page 60 that hysteresis is accompanied by a dissipation of energy within the substance. Suppose, for example, that our ball were being pulled up the hill by a string wound on a drum driven by an electric motor. In order to get it to the top we should have to supply a certain amount of electrical energy to the motor to enable it to do the work necessary to pull the ball uphill against gravity. If as soon as the ball started to move down the other side of the hill we switched the motor off and instead let it be turned by the unwinding string, the motor would behave as a dynamo, and, neglecting the effects of friction, the dynamo would generate electrical energy as the ball goes from C to D. If we were to reverse the motor we could by a similar process get the ball back to A without having used up any energy to do so. But if we left the motor running while the ball runs downhill we should not get this energy back and we should lose a certain amount of energy in moving the ball from A to E and thence back again to A. Now this is precisely what happens when we magnetize a ferromagnetic substance; we do not switch the field off at the beginning of every Barkhausen jump in the magnetization and we pay the price accordingly. Consequently we always have to supply energy to take a substance through a

Ferromagnetic Domains

hysteresis cycle, and this has to be supplied by the battery or generator or whatever we happen to be using to supply the current to magnetize the substance.

The energy lost per cubic centimetre of material we have already seen to be equal to the area closed by the hysteresis loop and this so-called hysteresis loss is a serious loss of electrical energy in alternating current machinery. The energy appears as heat in the material, and this is caused by the fact that every sudden jump of a domain wall reverses the magnetization locally, and by Faraday's Law this gives rise to an electromotive force, which in metals produces localized currents (eddy-currents on a small scale) which cause heating by virtue of the electrical resistance of the metal.

The total heating which occurs when a ferromagnetic substance is taken round a hysteresis loop is usually quite small, and it has been calculated that a piece of ordinary iron would have to undergo 4,000 complete cycles of magnetization in order to raise its temperature by one degree centigrade. Nevertheless careful experiments using highly sensitive apparatus has made it possible not only to measure the heating effect during one cycle but to find out those parts of the cycle where most of the heating takes place. Temperature changes of as little as one millionth of a degree have been measured in these experiments, which show that the energy losses are greatest where the hysteresis loop is steepest.

It will be clear from a look at the shape of the hysteresis loop that the area under the loop or the hysteresis loss is very roughly equal to the product of the coercive force and the remanent magnetization. We know at least qualitatively what the factors are that determine the coercive force. What are the factors that determine the remanence? The answer to this may be found from the knowledge that rotation out of the easy directions is a reversible process, so that if a substance is saturated and the field removed the spontaneous magnetization will spring back into an easy direction; presumably into that easy direction which lies nearest to the direction of the magnetic field. The remanent magnetization therefore depends largely on the number of easy directions and their orientation with respect to,

Magnetism

the magnetic field. For example, in a single crystal of iron magnetized along an easy direction the spontaneous magnetization remains in that direction when the field is removed, and consequently the remanent magnetization is one hundred per cent. At the other extreme limit, in a single crystal of cobalt magnetized in the difficult direction or, what is magnetically equivalent, a polycrystalline rod of nickel under tension, the easy directions are at right angles to the field and for these cases the remanence is zero. In normal polycrystalline substances, in which the crystal grains are directed at random, a calculation shows that the remanence should be about 85 per cent that of saturation if the crystal grains have a cubic structure with six or eight easy directions. If the crystals are not directed randomly but have a tendency to be aligned more along one direction than another there may be a preponderance of easy directions near the field direction and the remanence may be increased.* This alignment of crystal axes along a particular direction can often be achieved by making the material in sheet form and reducing its thickness by passing it through heavy rollers. Often the degree of alignment produced in this way is so great that the resulting substance has properties which closely resemble those of a single crystal.

Polycrystalline materials which are composed of crystallites in which there are only two directions of easy magnetization have a rather lower remanence, for of the two easy directions one must lie nearer the field than the other, so at remanence the magnetization of a domain lies in that one, whereas in the demagnetized condition as many domains must be magnetized in the opposite direction. Calculations show that in this case the remanence is just half the saturation magnetization. These values of the remanent magnetization 0·5 and 0·85 for the two types of material are representative of the values usually observed. Values outside these limits which are exhibited by single and quasi-single crystals can usually be satisfactorily explained.

Thus we see that the domain hypotheses, originally introduced as being necessary to explain the fact that a ferro-

* For similar reasons, *mutatis mutandis*, it may be decreased.

Ferromagnetic Domains

magnetic substance can exist in an unmagnetized condition, satisfactorily explains, at least in a qualitative manner, most of their observed magnetic characteristics. Moreover the domains may be seen and their growth and movement studied by means of the magnetic colloid technique. Apart from the obvious satisfaction of understanding how things behave as they do, the domain concept has proved singularly useful in one very important respect, for knowledge of domains – their shape and sizes and the factors that assist and impede their growth in magnetic fields – give us a means of controlling magnetic properties, and it is now possible to make magnetic materials whose properties are tailored to fit the appropriate function they are intended to perform. This aspect forms the subject of the following chapter.

CHAPTER 9

Magnetic Materials and their Applications

MAGNETIC materials are essential for the operation of many devices and machines used in modern industry. Without them the production and distribution of electricity would be virtually impossible and communication by telephone and radio would be limited to short distances only. Iron is by far the major constituent of all commonly used magnetic alloys, and some idea of the technical importance of magnetic materials may be gauged from the fact that several million tons of iron are produced annually for use as magnetic material.

Most of our present knowledge of the magnetic properties of the ferromagnetic element and their alloys stems from fundamental studies, often made on single crystals, some of which were carried out before the Second World War but which were greatly intensified after it. As a result we now know a good deal about the way the spontaneous magnetization of the ferromagnetic elements vary when they are mixed together to form an alloy or when other non-magnetic elements are added to them. Even more important from the technical aspect is that we also know how the magnetic anisotropy and magnetostriction depend upon the composition of a large number of alloys, and these factors as we have seen are decisive in determining their ease of magnetization in small magnetic fields. Moreover studies on ferromagnetic domains have contributed greatly to our understanding of the other factors which make for easy magnetization (for example, purification and freedom from mechanical stresses). As a result we can often predict with fair accuracy the magnetic properties of an alloy of any given composition provided its mechanical and thermal treatment are also known. Alternatively, given a required set of magnetic characteristics it is often possible to find a suitable alloy without too great a use of the method of trial and error. It is therefore easy to gain the impression that research into magnetic materials has

Magnetic Materials and their Applications

reached a stage at which it might well stop and the effort directed elsewhere. However there are many reasons why this is not so. In the first place there are still a number of factors which affect magnetic properties in a way we do not fully understand – an example is the fact that polycrystalline materials magnetize more easily if they are composed of large crystal grains than if the grains are small – and secondly the development of new materials is invariably followed by new applications which take advantage of new or improved magnetic properties, and this nearly always brings new facts to light which require fresh or further investigation. An example here is the application of ferromagnetic materials as memory devices in electronic computers (see page 162), which has greatly stimulated research into the reasons for the rectangular hysteresis loop, which is an essential requirement of these materials. Finally there is always the economic factor. Industries, particularly those producing magnetic machinery, do not necessarily want the best magnetic material there is, but the best at the price, and this usually means one in which the raw material is relatively cheap and which can be produced in quantity without expensive processing. On account of their relatively low cost the commonest magnetic materials are those containing the largest fraction of iron, and considerable effort is at present being made to improve their properties by means not involving the addition of another, possibly expensive alloying element. For similar reasons it is obviously important to see that when expensive metals are used in magnetic alloys this is done in the most economic manner. In general the most useful magnetic materials are either those which have a high permeability and are magnetized and demagnetized easily, or those which are permanent magnets and can be magnetized only with difficulty but retain their magnetization tenaciously. These two broad divisions are usually referred to as soft and hard magnetic materials respectively.

Soft magnetic materials are used extensively in all electrical equipment but by far the greatest quantity is used for the generation, transmission, and ultimate use of electrical energy, that is to say in the construction of dynamos, transformers, and electric

Magnetism

motors. Anyone who doubts that the production of these three commodities should annually consume nearly a million tons of iron should remember that, apart from the more obvious use of electric motors to drive industrial machinery, the electric motor is an essential part of numerous household appliances. All the electrical power needed to operate these – and consumption increases annually – has to be generated and applied through two transformers, for, although it is economically advantageous to transmit electrical power at high voltages and low current along relatively thin conducting wires, it is convenient to generate and consume the power at low voltages and high currents. The output voltage of the generating dynamo is thus stepped up by a transformer for transmission purposes and stepped down at the end of the transmission cable to a voltage at which it can conveniently be consumed. The requirements for a magnetic material for a transformer coil are somewhat similar to those required for electrical machinery such as dynamos and motors, and it is therefore necessary only to consider the chief aspects of the transformer problem.

As mentioned on page 61 a transformer operates on the principle of electromagnetic induction and consists of two coils usually wound over one another, so that if a changing current flows in one an electromotive force is induced in the other. If the current in the primary coil is alternating then there is an alternating e.m.f. set up in the secondary. The efficiency of a transformer is greatly increased by inserting a core of magnetic material inside the coils, for since the magnitude of the induced e.m.f. is proportional to the rate of charge of induction within the coil rather than rate of charge of field it is clear that for a given current in the primary the induced e.m.f. will be greater the higher the permeability of the core material. This is the first requirement therefore – a permeability as high as possible. Unfortunately it is not the only one. The current in the primary changes direction every hundredth of a second and consequently the material in a transformer coil is taken round a complete hysteresis cycle in twice this time, or at the rate of fifty hysteresis loops a second. Now we know that hysteresis means dissipation

Magnetic Materials and their Applications

of energy in the form of heat in the material, in the present case in the transformer core. Not only does this lead to a serious waste of electrical energy, but in certain cases the heating of the cores of charge transformers may be so great as to necessitate special cooling arrangements. Evidently the hysteresis loss of a suitable transformer core material must be made as low as possible if the efficiency of the transformer is to be high. Nor is

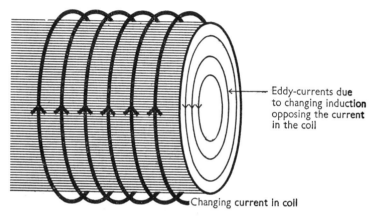

Figure 31. A cylindrical bar being magnetized by the current flowing in a coil wound round it. As long as the current is changing eddy-currents are set up within the bar in the opposite direction to the current in the coil, thereby decreasing the magnetic field at its centre.

this all, for the changing flux in this transformer core induces currents not only in the secondary winding but in itself. These are known as eddy-currents and their effects are two-fold. First, as Figure 31 shows, they always act, by Lenz's Law, in such a way as to oppose the change which produces them, which is to say that they always flow so as to create a magnetic field in opposition to the field produced by the primary current. In the case of a circular solid core these eddy currents flow round the circumference, and if the primary current is varying at the usual frequency of fifty cycles they can easily be of such strength that at the centre of the core there is no resultant magnetic field

153

Magnetism

acting at all and the inside of the core might just as well be removed. These eddy currents also produce heating of the core as any current flowing in a metal gives rise to heating effects. This represents an additional waste of energy which is undesirable. The eddy currents and their effects can be much reduced by increasing the electrical resistance of the substance and also by splitting up the core into sheets, each electrically insulated from one another.

Because of the enormous quantities of electrical energy used nowadays the annual magnetic losses in transformer cores and in electrical machinery is quite startling. It has been estimated that in the United States alone something like three hundred million dollars' worth of electrical energy is lost every year in this way, and the importance of magnetic losses has, not unnaturally, led to attempts to reduce them.

Before 1900 the material commonly employed was ordinary steel containing a very small amount of carbon. Apart from its very high losses this material suffered from a curious complaint known as magnetic ageing, now known to be due to the rather large number of impurities which it contained, and which manifested itself as a gradual decrease of permeability and an increase in the losses. The first big improvement came about 1900 with the discovery that small quantities of the element silicon added to iron brought about a remarkable improvement in magnetic properties. These are due to the increased electrical resistance, which lowers the eddy current losses, and an overall increase in permeability and decrease in hysteresis loss. This is known to be due to a reduction in the magnetic anisotropy of iron brought about by the addition of silicon. Moreover the presence of silicon brings about metallurgical changes which result in the almost complete elimination of magnetic ageing. The steady improvement in magnetic properties from 1930 are the just reward for the knowledge gathered from systematic fundamental investigations on ferromagnetic substances described in the previous chapter, and are due to the recognition of the deleterious effects of impurities and internal strains and the steps taken to eliminate them.

Magnetic Materials and their Applications

Another development leading to improved performance was also begun in the nineteen-thirties. Usually the material for transformer cores is obtained in sheet form by heating it in a furnace and rolling it while still hot, and this process is repeated until the required thickness* is attained. Reduction in thickness can however be effected by rolling the sheet while it is still cold. This causes the sheet to become rather brittle, and to restore some of its mechanical resilience the sheet has subsequently to be annealed. It is then found that the crystal grains in the sheet are no longer randomly orientated but show a strong tendency towards alignment with a [100] crystal axis in the direction of rolling (but with a [110] direction at right angles). By carefully controlling the extent of the reduction in thickness brought about by rolling and by suitable choice of annealing temperature almost perfect alignment of the crystal grains can be obtained in this way, and the sheet, although polycrystalline, has most of the desirable properties that one would normally associate with a single crystal.

In addition to the transformers used in what is called heavy current or power engineering there are others which are extensively used by the light current engineer. Here the transformer problem is rather different. The transformers involved are those used in radio and television receivers in the telecommunications industry. We may perhaps be more familiar with them as the connecting link between a high-quality gramophone pick-up and the amplifier, or between a microphone and tape recorder. In all cases the duty of the transformer is the same – to take through its primary coil the minute currents generated by such feeble electrical sources as microphones so as to generate a large e.m.f. in the secondary winding, which for this reason is wound with very many turns of wire. The primary currents are usually so feeble in this type of application that it is only the permeability at the very beginning of the magnetization curve or

---

* Usually between ·014 and ·030 inches, depending on its ultimate application.

Magnetism

the initial permeability* which really matters. The economic problem, too, is quite different, for these transformers are invariably small, usually no bigger than a matchbox, and so the cost of the raw material used in the manufacture of the magnetic core material is no longer so important a factor; indeed it may be quite insignificant in comparison with the cost of manufacture. Consequently the alloys used in magnetic materials usually contain appreciable quantities of nickel, in spite of the fact that it is much more expensive than iron.

The improvement in magnetic materials for low current transformers, which is to say the improvement in initial permeability, for this is the overriding factor, over the last sixty years

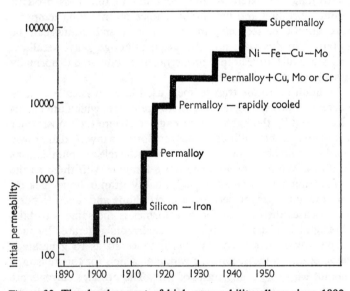

Figure 32. The development of high-permeability alloys since 1890. Note the compressed vertical scale.

* This can be taken to be 4π times the initial susceptibility in the present context. The initial susceptibilities of some of the alloys used in light engineering may reach three or four thousand, so that 4π times this is overwhelmingly greater than one, which may therefore be neglected in the expression $\mu = 1 + 4\pi\kappa$.

Magnetic Materials and their Applications

is shown in Figure 32. Nickel-iron alloys containing 35 to 100 per cent nickel are known as Permalloys, and they have the remarkable property that their magnetic characteristics are strongly dependent upon the way they are tempered. In the alloy containing 78 per cent nickel the highest permeabilities are attained after heating to a high temperature and then cooling as rapidly as possible. By adding small amounts of the elements copper, chromium, or molybdenum either separately or sometimes together, further substantial improvement in initial permeability may be obtained, and this is accompanied by an increase in the electrical resistance, thereby reducing eddy current losses in the material. The highest initial permeability is attained in a Permalloy, containing 5 per cent molybdenum, known as Supermalloy, which after suitable heat treatment attains an initial permeability of 120,000, which is just about one thousand times that of ordinary iron. The fact that its saturation magnetization, and those of the Permalloys in general, are rather less than half that of iron is of no consequence, since transformers containing these as core materials are always designed so that the core never reaches saturation.

The discovery of Supermalloy was announced in 1947, and it may be wondered why it has remained supreme for what, by modern standards, is a long time. Actually there is no reason to believe that the initial permeability of Supermalloy has not been surpassed in laboratories on isolated occasions, but it must be admitted that any substance possessing a significantly higher permeability would have little other than curiosity value. These high-permeability alloys are obtained only by exceedingly careful purification and relief of internal stresses. Unfortunately it is almost impossible to make a transformer with these high-permeability alloys as cores without re-introducing some strains. This has the effect of lowering the permeability to more normal values and we have every reason to believe that the higher the permeability a substance attains the more sensitive to the effects of strain it will become. Thus, although it is fairly certain that alloys can be made having a higher initial permeability than that of Supermalloy, the impetus to discover

Magnetism

them has decreased, and less effort is now being directed towards this goal.

Another important application of high permeability magnetic materials is in magnetic screening. Such screens have become essential in many types of elaborate communication equipment where instruments and sensitive devices must be protected from magnetic fields produced elsewhere in the apparatus. The effect of a magnetic screen is to divert the magnetic lines of force through the screen away from the object to be shielded. Theory shows that magnetic screening is never perfect unless the permeability of the screening material is infinitely great, and we can readily appreciate that the best screening material is that which has the highest permeability. Magnetic screens made from Permalloy are extensively used for screening the cathode-ray tubes in television receivers, thereby eliminating distortion of the picture due to the magnetic fields generated by external devices and components within the receiver. Efficient magnetic screening leads to greater compactness of design in all electrical apparatus, a fact which is partly responsible for the gradual reduction in size of television receivers and gramophone record players.

Soft magnetic materials are an essential constituent of the electromagnetic relay which frequently forms the heart of automatic and remote control systems. There are various types of relay but all operate on the principle that a piece of soft iron is attracted to the poles of an electromagnet. In a relay the soft iron is pivoted and held in position by springs. When a current is passed through the coil of the electromagnet the soft iron moves and in doing so operates one or more electrical contacts which can be used to switch currents on or off in other electrical systems. By using magnetic material of high permeability in the core, the electromagnet can be magnetized almost to saturation by a very small current in the coil, which is all that is required for its operation. On the other hand the force exerted by the electromagnet on the soft iron is quite large, large enough in fact to operate the heavy switches which are needed to start and stop large currents. The relay therefore acts as a kind of intermittently operating electric motor.

Magnetic Materials and their Applications

A group of relays connected together so that each influences the working of the other in the appropriate manner can be used to perform a number of remarkable tasks. First of all, a relay can be connected so that its switches control its own current, so that as soon as the current is started in the coil it is immediately switched off by the mechanical action of the relay. As soon as the current is switched off, however, the spring forces the soft iron back to its original position, thereby switching the current on again, and so the whole process is repeated indefinitely. This is the principle of the electric bell. It is however within the automatic telephone exchange that relays perform their most complicated tasks, and to do this a large automatic telephone exchange may contain between ten and forty thousand relays. It is not possible to explain the complicated manner in which they are arranged to do this, but some idea of the task involved may be gathered from the fact that not only do they have to select the telephone which is dialled from a possible number which maybe exceed fifteen thousand, they must also be able to give the correct 'dialling', ringing, and engaged tones, and must be so organized that subscribers get the right number without interference from the hundreds of other subscribers who may be dialling at the same time.

There is another group of materials which on account of their special properties have applications which are different in principle from those already mentioned. These are the materials which exhibit a large magnetostriction and they are commonly used to convert electric oscillations into mechanical ones and vice versa. Nickel is still one of the most widely used of these materials, although others are now being developed which may supersede it. The principle of the magnetostriction effect has been mentioned in Chapter 4. Application of a field to a ferromagnetic substance causes it to expand or contract. If the field is an alternating one the expansions and contractions will alternate, and the substance will be set into oscillation and will emit a sound just as a bell does when it is struck. Actually the nature of the oscillations which a magnetostrictive substance undergoes is more akin the oscillation which can be set up in a wine glass by rubbing a moistened finger round the rim. The sound

Magnetism

produced by this type of vibration is much higher in pitch (i.e. corresponds to more vibrations a second) than if the glass is merely gently struck, and with ferromagnetic metals it is easy to arrange that they vibrate in such a way that the sound waves they emit are too high in pitch to be heard by human beings. These high-frequency, inaudible sound waves are usually known as ultrasonic waves and are not only of importance as a general investigational tool in science, but have interesting and useful applications as well.

One of the best-known applications of ultrasonics is in their use for the detection and location of underwater objects. High-frequency sound waves (ordinary sound waves will not do because they spread out too much and in consequence only a minute fraction of the original wave would be received after reflection) usually at a frequency which varies between ten and fifteen thousand vibrations per second, are generated in short bursts by the magnetostrictive device, and are directed towards the seabed. The depth is determined by the time interval between transmitting the burst and receiving it back after reflection, the burst of ultrasonic waves usually being recorded on a moving chart. The original application of depth sounding has now been extended to the detection and location of various underwater objects such as fish shoals and wreckage. Great use was made of this device during the war to detect submarines, torpedoes, and mines. So important did these ultrasonic methods of location become that in Japan, where the shortage of nickel was very acute, a good deal of research went into developing a suitable alternative. It was found that by adding aluminium to iron an alloy could be made with magnetostrictive properties very similar to those of nickel, and this alloy was used as a substitute while supplies of nickel were short.

It is also possible to construct an electrical memory device using magnetostrictive action, which is somewhat simpler in principle than the memory array using ferromagnetic cores which we shall encounter later. This consists essentially of a long wire, usually of nickel, magnetized at each end by a small permanent magnet (see Figure 33). Two small coils are wound

Magnetic Materials and their Applications

round each end and the information, in the form of electrical pulses, is sent through the first coil. Each pulse of current magnetizes the wire momentarily, and this magnetization pulse produces a mechanical stretching pulse which travels down the wire just in the same way that the movement of a clothes line flicked at one end with the hand is conveyed down to the other end. On arriving at the other coil the stretching pulse changes the magnetization in the wire (by the inverse magnetostriction effect) and a pulse of current is set up in this coil which, if everything is correct, should be identical with the original current pulse in the first coil. Several factors combine to prevent this from being exactly so, and the received pulse is amplified by an amplifier, trimmed up in shape, if necessary, and sent back to the first coil, when the whole sequence of operations begins again. Thus the information, in the form of electrical pulses, is kept circulating round the system and is stored until

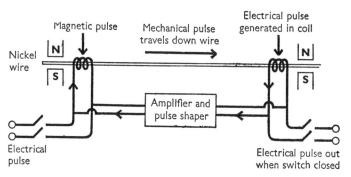

Figure 33. A simple magnetic storage device. The information in the form of electrical pulses is used to generate a magnetic field pulse which, by the magnetostrictive effect, gives rise to a mechanical pulse. This travels down the wire at the speed of sound in the wire (about 4×10^5 cm./sec.). On arriving at the end the mechanical pulse changes the magnetization in the wire by the inverse magnetostriction effect and induces an e.m.f. in the coil. Ideally this pulse should be identical with the original, but some distortion and loss of strength always occurs and this is put right by the amplifier which also re-shapes the pulse. The small permanent magnets at each end of the wire serve to keep it in the optimum magnetized state.

Magnetism

the time when the information is needed. Then a system of switches stops it from completing a closed circuit but instead leads the pulses elsewhere where they may perform their proper task. This type of storage element is used in automatic telephone exchanges where the information to be stored, from the dialling system for example, is usually fairly simple in content and can be accommodated without much difficulty by this means. The total number of pulses which can be stored in this way is limited by the length of the wire which may be used and usually it is not possible for this to exceed a few feet.

It has been mentioned on page 146 that certain alloys can be made to have hysteresis loops so steep-sided as to be almost rectangular in shape. There are a variety of ways of achieving this state, but one of the simplest ways is to heat the alloy and allow it to cool in a magnetic field. Not all alloys respond to this treatment, but a good many do. The effect is to create a direction of easy magnetization in the direction in which the field was applied, and the material then possesses many of the properties of a single crystal. Magnetic materials having a rectangular hysteresis loop are rapidly coming into use in memory units which store the information obtained during a certain sequence of operations of an electronic computer until such time that the machine is ready to use the stored information for the next operation.

It is well known that any information can be obtained from a series of questions requiring the answer yes or no, this being the basis of a large number of parlour games. The information is stored by a computer in a similar manner, as a number of 'yes' or 'no' bits. Since there are no alternatives to 'yes' and 'no' all that is needed to make a memory which will store one bit of information is a device which can exist in two possible states such as a switch which can come to rest in the 'up' or 'down' position only. A ferromagnetic substance satisfies this requirement since, once it has been magnetized, it can exist only in one of the two states of remanence, one of which can be called '0', the other '1', corresponding to 'no' or 'yes'. The way in which ferromagnetic cores are used as memory devices is shown in Figure 34. The cores are arranged in a rectangular array and

Magnetic Materials and their Applications

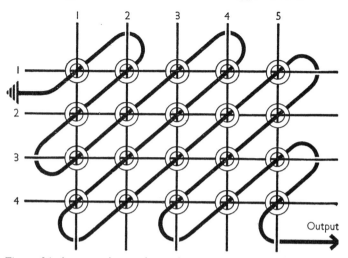

Figure 34. A magnetic matrix used as a memory or storage device.

each has three separate coils wound on it connected as shown. Information is the form of a series of pulses – a positive pulse corresponding to '1' and a negative one to '0' – is fed into the memory and stored there until required. The method of writing a '1' into a particular core, for example the one labelled 4·3 in the fourth column and third row, is to pass a pulse of current through the fourth vertical wire and at the same time an identical pulse of current through the third horizontal wire. If each current pulse produces a field which is slightly less than the coercive force of each core no effect will be produced by a single pulse alone, but core 4·3 is acted upon by two identical pulses which together produce a field greater than the coercive force.

Thus the magnetization in core 4·3 is reversed while that in all the other cores remains unchanged. In order to 'read' this '1' at a later time, current pulses are again passed through the fourth vertical and third horizontal wires but this time in the opposite direction. The magnetization in core 4·3 is then reversed and induces an electromotive force in the third winding which threads all the cores, indicating that a '1' had been

Magnetism

stored in core 4·3. It will now be clear why the hysteresis loop of a magnetic material suitable for this application should be rectangular. If a '0' is stored in a particular core it is desirable that there be no output pulse, i.e. no induced e.m.f., when reading. In a normal material there would be one corresponding to the change of magnetization within the core from remanence to somewhere near saturation, and unless this can be made small in comparison with the pulse produced by change from remanence positive to remanence negative the distinction between a '1' pulse and a '0' pulse will become lost and the device will no longer act as a memory. This is achieved by using materials with rectangular hysteresis loops in which the remanence is at least nine-tenths of saturation, so that the unwanted pulses produced during reading are kept very small.

This so-called magnetic memory array was first devised at a time when quite large numbers of rectangular loop materials were known. However, the interesting thing is that its invention did not merely provide one more application for materials already in existence. It stimulated new research into the reasons for rectangularity of the hysteresis loop and it led to the discovery of rectangular loop characteristics in materials where one would not expect to find them and, in this way, to further unsolved problems. In addition it raised to a new status of importance a completely new factor which had previously not been investigated, that of speed of response.

Electronic computers are able to perform in a short time calculations which would take a human being months or years. The speed of calculation of the usual types of computer, although remarkable by human standards, is however not great enough. In the first place computers are large pieces of equipment costing a great deal of money to manufacture, maintain, and operate. Consequently these machines are economic only as long as they perform quickly, and the more rapidly they perform the more efficient and economical they become. Furthermore increasing the speed of operation of computers increases the scope of their application, for in spite of their agility there are problems which are so complex that even

Magnetic Materials and their Applications

with present-day computers the time required for their solution would be prohibitively long. In consequence one of the chief objectives in the development of new computers is increased speed of operation.

One of the factors which limits the speed of operation of a computer is the rapidity with which the information can be stored and read out again, and one of the factors which limits this is the speed with which a ferromagnetic core can be 'switched' from one remanent state to another. What determines the time required to 'switch' a ferromagnetic core and what steps can be taken to decrease it? What rectangular-loop material has the smallest 'switching time'? These are the questions which the designers of memory arrays were asking in the early nineteen-fifties, and to which there were no ready answers. Thus began a whole new aspect of research in magnetism. The problem is no longer merely one of what happens to a magnetic material when it is magnetized; it is necessary to know how quickly this happens and why.

Investigations showed that in metallic ferromagnetics the chief obstacle to rapid switching is provided by the eddy currents set up in the core itself. These can be reduced by making the cores out of thin strips of magnetic material often coiled into circular cores like clock springs. In order to reduce eddy currents to negligible proportions the thickness of these strips is often reduced to as little as 3×10^{-4} cm. Such strips are produced by successive rolling of thicker material and the skill and precision required to make metal of this thickness (thinner than the pages of this book), with definite magnetic characteristics, is immense. Attempts are now being made to use even thinner material by depositing thin films of iron by evaporation in vacuo. Unfortunately the thinnest strips show that when eddy currents have been eliminated entirely, the time taken for switching, usually about one millionth of a second, is determined by another factor which still remains something of a mystery. So there is more scope for further investigation and research.

Another way of avoiding eddy currents is to use a material of very high electrical resistance. The ferrites are examples of

Magnetism

these and though we are not due to meet them for a while it is only fair to add that their use in memory arrays is extensive.

The computer problem has been discussed here at unusually great length not only because it provides one of the more recent applications of magnetic materials but because it illustrates very clearly the very close connexion and interrelation between pure research and the application of its discoveries. Computers promise great things in the service of mankind, for they are likely to cover a far greater range of application than the mere solution of the problems of large business organizations. Scientific and engineering problems which, although not difficult in principle, become intractable in practice owing to the large number of factors which have to be taken into account, can be solved and the automatic translation of books has now become technically feasible. Magnetic materials play an essential part in the work of these automatic brains.

CHAPTER 10

The Development of the Permanent Magnet

IMPROVEMENTS in magnetic materials for use as permanent magnets during the last sixty years have been no less remarkable than those in soft materials. Permanent magnets have a longer ancestry than any other form of magnetic material, and it is appropriate to devote a whole chapter to them. This is desirable in many ways, for the application of permanent magnets are numerous and the history of the modern permanent magnet is a remarkable record of the very close interplay between purely scientific discoveries on the one hand and their technological application on the other.

The earliest magnetic substance known to man was a permanent magnet – lodestone – the magnetic oxide of iron. And indeed lodestone remained for at least two thousand years the only permanent magnet and the only means by which iron could be magnetized. The first artificial permanent magnets were iron needles which had been 'touched' by the lodestone, and it is possible that the earliest forms of the compass made use of a needle magnetized in this way.

The first systematic study of the magnet was that undertaken by William Gilbert, the results of which he published in his famous *De magnete* in 1600. One of Gilbert's first duties was to demolish the numerous superstitions that were associated with the lodestone. There were many reports of the curative powers of the lodestone, though some held that lodestone was a poison. Others believed that the lodestone acted only by night. An example of the traditional beliefs refuted by Gilbert is that the lodestone lost its attractive power in the presence of certain special substances, notably garlic, goats' blood, and diamond. Gilbert's method of investigation, unlike the superstitious speculations of most of his contemporaries, was simple and direct, and one which would be appreciated by a scientist of today. He put seventy-five diamonds around a lodestone and

Magnetism

then proceeded to demonstrate that its attractive powers were unimpaired.

The most powerful magnets of Gilbert's time were still lodestones. He described how to increase their power to attract iron by 'arming' them, that is by attaching soft-iron caps to each end. In the seventeenth century armed lodestones were much in demand, chiefly for their curiosity value, and for this purpose they were usually improved by squaring the ends and applying suitably shaped iron plates. Often they were mounted in silver and suitably fashioned for the wealthy amateurs who were the only people who could afford to buy them. Gilbert, however, was much concerned with making permanent magnets, or artificial magnets as he called them. He was quick to observe that the best magnets were those formed from 'iron made hard' and that soft iron, however efficacious in arming lodestone, will not by itself make a good magnet. It is interesting to note that 'iron made hard', usually by adding carbon and thereby making steel, remained the standard permanent magnet material for nearly three hundred years.

Gilbert gave three ways of imparting permanent magnetism to iron and steel. The first was the age-old method of 'touching' with a single lodestone, which was to be drawn from the centre of the needle to be magnetized to its end. His second observation was that iron wires became magnetized if drawn in a north-south direction, but not if drawn in an east-west direction. We would now recognize this as magnetization by the combined effect of a magnetic field, that of the earth, and the effect of mechanical stress, for the magnetostriction of iron is positive in weak fields and so the effect of tension on an iron wire is to render it more easily magnetizable by the earth's field. The third method arises from the fact that a red-hot iron bar left to cool in the direction of the earth's field becomes permanently magnetized. A similar effect can be produced by forging an iron bar or by hammering it when cold if it is similarly placed (Figure 35). This is the mechanism responsible for the magnetization acquired by ships' hulls when the plates are riveted or welded together.

During the eighteenth century the making of powerful per-

The Development of the Permanent Magnet

Figure 35. Forging an iron bar placed along the direction of the earth's magnetic field – an illustration from Gilbert's *De magnete*. Note: Septentrio, north; Auster, south.

manent magnets claimed the attention of several eminent men of science, notably Garvin Knight (1713–72), an Englishman whose magnets became renowned over the whole of Europe for their strength and tenacity. Oersted's discovery of the magnetic effect of an electric current rapidly led to the development of the electromagnet, and this together with the decline in the wealthy amateur, who in the eighteenth century was always a customer for scientific curiosities, seriously decreased the incentive to pursue studies of permanent magnets. There was no industrial demand for them, and by 1860 or thereabouts serious investigation of permanent magnets had ceased to be carried out. Scientific interest in their potentialities seems to have been roused by the work of J. C. Jamin (1818–86), a French physicist best known for his work on optical interference, who in 1873

Magnetism

constructed what he claimed to be the most powerful permanent magnet in the world. It weighed nearly a hundredweight and was capable of supporting a load ten times its own weight. Jamin's work is important; he chose his steel with great care and went systematically into such questions of design as the relative volume of the pole pieces and the best material for their construction. Jamin's was thus the first permanent magnet to be constructed according to a proper scientific design.

By about this time the need for improved permanent magnets was beginning to be appreciated, both for purely scientific work and for its applications. They were needed in order to detect the very feeble signals received from the recently-laid telegraph cables (the first successful submarine cable across the Atlantic had been laid in 1866). The newly invented telephone required for its successful operation a strong permanent magnet. Electricity was beginning to be generated in quantity, and among the uses to which it was being put were some requiring the service of a strong permanent magnet. On the purely scientific side the technique of electrical and magnetic measurements had improved considerably, and there arose a real need for measuring instruments, such as ammeters and voltmeters, which were more sensitive and at the same time more accurate than those which were in use at the time. As a result research began to be initiated into ways and means of improving permanent magnets. Careful design helps a great deal, but it was known that hardened steel made a much better permanent magnet than unhardened steel or soft iron, and so the question was asked whether there might not be other new materials out of which even better magnets might be made. In this way the search for new and better permanent magnet materials was begun, a search which has led to permanent magnets which today are as superior to those of sixty years ago as the magnet of 1900 was to the armed lodestone of Gilbert's day, three hundred years previously.

What are the particular magnetic characteristics which go to make a good permanent magnet? Obviously the magnet must retain as much of its magnetism as possible when the magnetizing force is removed, which is another way of saying that the

The Development of the Permanent Magnet

remanent magnetization must be as high as possible. This cannot be the only factor however, for as we saw in Chapter 8 the remanent magnetization of most substances is about half the saturation value and so should be roughly the same for all types of iron whether hardened or not. In fact the remanent magnetization of hard steel is not greatly different from that of soft iron. Some other factor must be operative as well, and to find what this is we must consider a little more carefully the way a permanent magnet is used.

The attracting power of a magnet is caused by the magnetic field it produces. Now the magnetic field due to a bar magnet is proportional to the number of magnetic poles per square centimetre at the end faces of the bar, and therefore to the intensity of magnetization within it. The magnetic lines of force of this field begin on a north pole and end on a south pole both outside and inside the magnet. The field inside the magnet therefore is in a direction opposite to that of the magnetization within it and thus tends to reduce its value. For this reason the field due to a permanent magnet, or for that matter to any magnet with end faces on which magnetic poles can appear, is sometimes known as a demagnetizing field. The demagnetizing field, being proportional to the intensity of magnetization, must be equal to some constant factor multiplied by it. This constant factor is known as the demagnetizing factor, and this in turn depends only on the shape of the magnet or magnetized body. It is zero for an infinitely long cylinder or for a closed ring with no faces upon which poles may be formed. It is $4\pi/3$ for a sphere and 4π for a thin plate magnetized at right angles to the surface. A bar magnet the shape and size of an ordinary lead pencil would have a demagnetizing factor of about one twentieth, so that if the magnet were made of iron the demagnetizing field at remanence would be one twentieth of the remanent magnetization (which for iron would be about 800) or about 40 oersteds; this would be sufficient to reduce its magnetization to zero. Of course it is a little more complicated than this, since any decrease in the intensity of magnetization is automatically accompanied by a decrease in the demagnetizing field and this has a kind of compensating effect, but the essential

Magnetism

point is that we must not overlook the fact that any permanent magnet tends to demagnetize itself to a certain extent, and that the best magnet for any particular application is one which is least affected by its own demagnetizing field. We can find this out by looking at that part of the magnet's hysteresis curve which shows the way in which the magnetization is decreased by a reverse field. This so-called demagnetizing quadrant is shown in Figure 36 for two materials A and B of different coercive force. Suppose for the sake of definiteness we decide

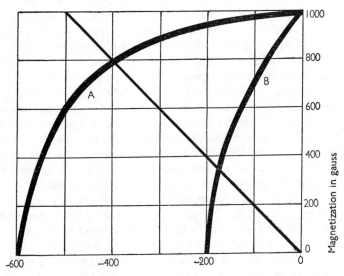

Figure 36. Permanent magnet design. The field produced by a permanent magnet is given by the point of intersection of the magnetization curve in the demagnetizing quadrant (between remanence and coercive force) and the line $H = -NI$, where N is the demagnetization factor of the magnet. In the example $N = \frac{1}{2}$ for two magnets made of materials with different magnetic characteristics. (A) $I_R = 1000$ gauss, $Hc = 600$ oersteds. (B) $I_R = 1000$ gauss. $Hc = 200$ oersteds. The field produced by magnet A is 400 oersteds, that of B about 170 oersteds, so for a magnet with $N = \frac{1}{2}$, A is about three times as good as B.

The Development of the Permanent Magnet

to make cylindrical bar magnets out of A and B whose lengths are four times their diameters (the actual size does not matter, it is only the relative dimensions that count). For such a rod the demagnetizing factor is about one half. The field inside the bar, and this is equal to the field just outside its faces too, will always be one half of the intensity of magnetization, I, of the bar. For a given demagnetization curve there is only one value of I which is possible for this value of the field, namely where the curve intersects the line $H = -\tfrac{1}{2}I$. For magnet A, whose

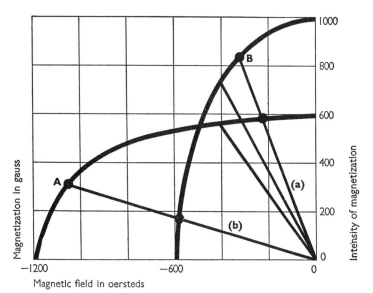

Figure 37. Permanent magnet design. The graph shows demagnetization characteristics of two materials: (A) $I_R = 600$ gauss, $H_c = 1200$ oersteds. (B) $I_R = 1000$ gauss, $H_c = 600$ oersteds. For a long thin magnet the demagnetization factor, N, is small and the line $H = -NI$, marked (a) in the diagram, intersects the two curves A and B at $H = 250$ and $H = 340$ oersteds respectively. In this case magnet B is superior to A. For a shorter magnet, N is greater and the line $H = -NI$, marked (b), is more horizontal. It intersects the two demagnetization curves at $H = 1060$ and $H = 590$ oersteds and in this case magnet A is superior to B.

coercive force is 600 oersteds, this occurs at $I = 800$ gauss and $H = 400$ oersteds. With magnet B, whose coercive force is 200 oersteds, the corresponding values are approximately $I = 170$ gauss, $H = 170$ oersteds, so that magnet A is actually more than twice times as strong as magnet B despite the fact that their remanences are identical. Thus a good permanent magnet is one with a sufficiently high coercive force to resist its own demagnetizing effect. This is of great importance, and in fact the great improvements in permanent magnet materials over the past sixty years are almost entirely the outcome of increases in their coercive forces. This does not mean that the remanent magnetization is unimportant, as a glance at Figure 37 will show. This gives demagnetization characteristics for two materials A and B with remanences 600 and 1,000 gauss and coercive forces 1,200 and 600 oersteds respectively. For a short thick magnet with large demagnetizing factor magnet A is superior, while for a longer thinner one having smaller demagnetizing factor B is the better material. Thus we see that the shape of the magnet, which may be determined by the duty it is expected to perform or the space it is allowed to occupy in a certain piece of equipment, determines the choice of magnetic material out of which it is to be constructed.

Another feature of permanent magnet alloys can be seen from a consideration of these two alloys A and B. Suppose it is required to make a permanent magnet to produce a magnetic field of pre-determined strength, say, 400 oersteds. This is a situation frequently encountered in the design of electrical apparatus. Figure 37 shows that either material will do work satisfactorily but that the slope of the curve $H - NI$ for material A (high Hc, low I_R) is larger than that for material B. This means that the demagnetizing factor of material A is larger than that of B which in turn means that if the two magnets are of the same cross-section magnet A is shorter than B. Thus a high remanence – low coercive force magnet can often advantageously be replaced by a lower remanence – high coercive force magnet of smaller size, thereby effecting considerable saving in space and in cost.

We have seen that neither the remanence nor the coercive

The Development of the Permanent Magnet

force by themselves determine the quality of a permanent magnet material and moreover that the same magnetic field can be produced by two different magnets operating at different parts of their demagnetization curves. Is there therefore any single factor which gives an indication of the overall quality of a permanent magnet material? Engineers prefer to work in terms of the magnetic induction B, equal to $H + 4\pi I$, instead of the intensity of magnetization, because it turns out that the field due to a permanent magnet depends upon B rather than I, and design figures are usually worked out in terms of B. A useful figure of merit is the maximum value of the product $B \times H$ attained in the demagnetizing quadrant of the hysteresis loop. It can be shown that this product, known as the energy product, is proportional to the square of the magnetic field which a well-designed magnet of a given volume can produce in the gap between its poles. Typical figures for a good modern permanent magnet are: remanent induction* $B_R = 10,000$ gauss, coercive force $Hc = 500$ oersteds, energy product 2×10^6 gauss-oersteds.

We now know what constitutes a good permanent magnet material. How can this be achieved in practice? Here our knowledge of domains and the movement of domain walls is of great assistance. From it we know the important factors which control the impedance to wall movement and thus the high coercive force, and in consequence we can nowadays search for new high-coercivity alloys in a scientific and systematic manner. Broadly speaking we know from the previous chapter that the factors that control the coercive force and the initial permeability are basically the same and any change in conditions which increases the one is likely to decrease the other. Thus a ferromagnetic substance may be given a high coercive force by deliberate introduction of internal stress, by hammering, repeated bending, or by drawing, for example. The coercive force of nickel can be increased from about one oersted to

---

* In the remainder of this chapter we shall use induction B instead of intensity of magnetization in conformity with standard practice. At remanence $H = 0$, and so the remanent induction is exactly 4π times the remanent magnetization.

about thirty in this way and at the same time the initial permeability drops from 300 to about 10. Likewise the impurities which are so detrimental to the achievement of a high permeability are often vital for the attainment of large coercive forces. Unfortunately these facts were not known to the early makers of permanent magnet material, whose alloys were made by cookery book methods, and improvements were made by blind groping or as a result of guesswork or sheer intuition. It is hardly surprising that progress was slow, and in fact it was not until about 1930 that really significant advances were made. Only one guiding principle seems to have been known, which is that any metal which is mechanically hard is likely to be magnetically hard as well. (The converse of this is not true.) Annealed nickel for example is not only magnetically soft but is so soft mechanically that it is almost impossible to handle it without bending it permanently. After drawing, or any other form of severe mechanical treatment, it becomes extremely hard, and this is accompanied by an increase in magnetic hardness as mentioned previously. We now know that mechanical hardness in metals and alloys is strongly bound up with the presence of internal stress within the material, and since these also impede the movement of domain walls, magnetic and mechanical hardness are likely to go hand in hand. The internal stresses which can be put into a metal by external work are quite small, and the hardest metals are those alloys in which the internal stresses arise from the misfit between its two constituent metals. The hardness of quenched steel for example is due to small isolated regions of the alloy iron carbide Fe_3C, which forms a crystal lattice slightly different in size from that of iron, and consequently cannot fit in without straining it severely. Not all alloys behave in this way but those that do are usually extremely hard, often to a degree which leads to brittleness. If they are magnetic these are the alloys in which the best permanent magnet qualities are likely to be found.

Thus the earliest permanent magnet alloys were glass-hard tool steels containing about 1 per cent carbon and having coercive forces of up to 50 oersteds. As early as 1855 alloying elements were added, especially tungsten, to improve their

The Development of the Permanent Magnet

quality, and in this way coercivities as high as 80 oersteds were attained. As permanent magnets these alloys were unsatisfactory in several respects. Their properties were extremely sensitive to changes in temperature and to mechanical shocks and vibration. In addition they suffered from the same ageing troubles that bedevilled the early transformer steels. These are serious defects. A permanent magnet in an aircraft for example is likely to undergo extremes of temperature and be subject to constant and intense vibration, and it is essential, if it is performing some vital task, that it maintains its original properties for as long as it is used. Permanent magnets are extensively used to provide the magnetic field in moving coil ammeters and voltmeters. The sensitivity of these instruments depends on the magnetic field in which the moving coil is placed, and so it is essential, if the instrument is to maintain its accuracy, that the field of the magnet remains constant. These effects can all be overcome to a certain extent, but, together with the fact that even the best cobalt steel does not make a really strong magnet, they were sufficient to hinder the widespread adoption of permanent magnets in numerous applications for which they were potentially suited.

In 1917 a considerable improvement was made by the discovery of the Japanese physicists K. Honda and T. Takei that the coercive force of tungsten steel could be vastly increased by the addition of cobalt. With 35 per cent cobalt the coercive force reaches 240 oersteds and the energy product about one million (10^6), which is about five times as great as that obtainable with the best carbon steels. At the time these alloys were expensive owing to the high cost of cobalt. They are still extensively used as they are mechanically stronger than the newer permanent magnet alloys and can be produced in complex shapes without fear of breakage.

The situation changed almost overnight as a result of a chance discovery, once again in Japan, in 1932. In the course of an investigation into certain corrosion-resistant alloys I. Mishima discovered that certain alloys containing iron, nickel, and aluminium, after suitable heat-treatment, developed coercivities as high as 700 oersteds. Two years later it was discovered

Magnetism

that improvements could be obtained by adding cobalt and copper, and the resulting series of alloys are usually known collectively as 'Alnico'. Energy products as high as $1 \cdot 7 \times 10^6$ were obtained, representing a three-fold increase in the field obtained from an Alnico permanent magnet over that available from the best cobalt-steel. Besides their magnetic superiority over all previous materials the new alloys were found to have other important properties. They are rustless and relatively free from corrosion. They are virtually unaffected by vibration and by temperature. In addition they do not exhibit the ageing characteristics of the early materials. Finally, they are about half the price of the best cobalt steel.

The impact of the introduction of these new alloys on everyday life was rapidly felt. Many people will be able to remember the excitement which accompanied the introduction of the first successful moving-coil loudspeaker with the injunction to 'listen to the bass'. The superiority of this form of loudspeaker over those commonly used in the early days of wireless had long been recognized, but the principle had not been put into practice owing to the very low efficiency which arose from the relative feebleness of the early permanent magnets. Another immediate consequence was the replacement of cobalt-steel magnets in certain applications by Alnico magnets of considerably smaller size. Thus the old-fashioned 'handbell' type of telephone earpiece, which was made in this form not for convenience in holding, but to accommodate a large horseshoe magnet, was replaced by the smaller earpiece combined with microphone which is now universally adopted in telephone handsets. The widespread adoption of dynamo lighting for bicycle headlamps which came about towards the end of the nineteen-thirties is another simple example of the results of the reduction in size made possible by the new magnets.

In 1938 it was found that by cooling these alloys in the presence of a magnetic field the remanent magnetization can be increased. The coercive force is unaffected by this treatment but the demagnetization curve is made more square and as a result the energy product is increased. These alloys, which may contain small amounts of titanium and niobium, are known

The Development of the Permanent Magnet

under various trade names – Alcomax, Ticonal, Alnico V and VI, and Hycomax – depending on the country of origin. Their remanences are between 10,000 and 12,000 gauss, coercive forces from 600 to 750 oersteds, and energy products up to 6×10^6 gauss-oersteds; most good-quality magnets are nowadays made from these materials.

Characteristics of Some Permanent Magnet Materials

Name	Typical Composition (weight per cent)	Coercive Force H_c (oersteds)	Remanence B_R (gauss)	Energy Product (BH) max.
HARD STEEL	99 Fe, 1 C	50	9000	$0 \cdot 2 \times 10^6$
COBALT STEEL	35 Co	250	9000	$0 \cdot 95 \times 10^6$
ALNICO	18 Ni, 10 Al, 12 Co, 6 Cu, 54 Fe	560	7250	$1 \cdot 7 \times 10^6$
ALCOMAX III	14 Ni, 8 Al, 24 Co, 3 Cu, 1 Nb, 50 Fe	670	12500	$5 \cdot 0 \times 10^6$
PLATINUM COBALT	77 Pt, 23 Co	4000	6000	$9 \cdot 0 \times 10^6$
VECTOLITE	$FeOFe_2O_3 +$ $CoOFe_2O_3$	900	1600	$0 \cdot 5 \times 10^6$
FERROX-DURE	$BaFe_{12}O_{19}$	2000	3500	3×10^6
BISMANOL	MnBi	3400	4300	$4 \cdot 3 \times 10^6$

The precise reason why these alloys possess such high coercive forces is still something of a mystery. It is known that the maximum coercivities are developed only after special tempering. It is also known that the effect of this tempering is to cause the alloy to change from a homogenous alloy into one consisting of two so-called phases, two distinct alloys of different composition inextricably intermingled so as to form a coherent alloy. One of these phases is known to consist of

Magnetism

the alloy Fe-Co, dispersed in the form of tiny islands in a sea of the other phase. This has two effects. The crystal lattice of these islands is not quite the same size as that of the other phase and the two can become coherent only if the latter is considerably strained. The Alnicos are in fact extremely hard mechanically and also very brittle. Furthermore the spontaneous magnetization of the two phases is different so that the material is magnetically inhomogeneous. Both factors will be recognized as being conducive to high coercive forces. Just how they give rise to coercivities of the magnitude found in the best permanent magnet material is something upon which universal agreement has not yet been reached.

The Alnico type alloys have one or two disadvantages which, though not often serious, have led to attempts to find other materials which would in certain applications be superior. They are brittle and have to be cast before being ground into their final shape. Furthermore there seems to be a limit of about 800 oersteds to the coercivity with which they can be endowed, and consequently in applications requiring a thick squat magnet they are of limited use. In this respect their properties are easily surpassed by an alloy consisting of 77 per cent platinum and 23 per cent cobalt. This remarkable alloy has a coercive force of 4000 oersteds and an energy product of 9×10^6, but its high cost seems likely to prevent it from coming into widespread use and it seems as though we must look elsewhere for new material.

One promising line of inquiry was approached in the following way: In order to make a substance magnetically hard one seeks ways and means of preventing domain walls from moving easily. Since it appears that domain walls can never be completely prevented from moving, why not remove them completely? Magnetization could then occur only by rotation of direction of the spontaneous magnetization, and we know this to be a 'hard' process from single crystal studies, very hard indeed in the case of crystals of cobalt type. Quite simple calculations showed that if there were no domain walls coercive forces would indeed be very great, and there remained only the problem of eliminating domain walls. This is actu-

The Development of the Permanent Magnet

ally easier to achieve than might at first be imagined, for it turns out that a ferromagnetic particle has to be greater than a certain size before domains will form; if it is made smaller than this there will be no domain walls. Such particles will behave as a single domain always spontaneously magnetized to saturation, and in a magnetic field the spontaneous magnetization will rotate as a whole and a very large field is necessary to reverse its direction.

It is easy to see why a small particle should be a single domain. Domains are formed to reduce the demagnetizing energy of a ferromagnetic body. This energy is proportional to the volume of the body, and for ordinary sizes the reduction of the demagnetizing energy by even a single domain wall is considerable and more than offsets the increase in energy due to the formation of a wall, which is proportional to its area. In the case of very small particles this may no longer be true, since with very small volumes the demagnetizing energy may already be smaller than the energy required to form one domain wall. Such a particle will exhibit no tendency to divide into domains, but will in fact always consist of a single domain. The critical size below which a particle becomes a single domain depends upon a number of factors including the spontaneous magnetization and the shape of the particle. For an iron sphere the critical radius is calculated to be about 10^{-6} cm.; for cobalt it is slightly larger. The magnetic field necessary to reverse the magnetization in one of these single domains depends more than anything else upon its shape, being least for a sphere and greatest for an elongated ellipsoid. The theory of the magnetic behaviour of an assembly of single domains was worked out in great detail by Stoner and Wohlfarth in England shortly after the Second World War.

The first suggestion that an aggregate of very small magnetic particles would have a coercive force much greater than that of the same material in bulk had come eight years earlier from a German geologist who had noticed that certain magnetic rocks possessed unusually large coercivities when the magnetic constituent was present in finely dispersed form. About 1941 news came from France of the successful manufacture of a new

Magnetism

material made from finely powdered iron and having remarkably good permanent magnet properties. Even today it is not known who was primarily responsible for this work, for France was at the time under enemy occupation and it was carried out under conditions of extreme secrecy. Possibly, too, the claims made for these materials became somewhat exaggerated, as the news came by various devious routes, and these so-called micropowder magnets, although commercially produced after the war, never seem to have had the success which was at one time expected of them. Much more recently however the problem was taken up again, and in 1956 magnets made of a finely powdered alloy of iron and cobalt were prepared in the laboratories of the General Electric Corporation in the U.S.A. with a coercive force of 1000 oersteds and an energy product of 5×10^6. In this material the particles are elongated and are aligned in the direction in which it is to be magnetized. Permanent magnets have also been made of finely divided powder of barium ferrite,* known commercially as Ferroxdure, of cobalt-iron ferrite (Vectolite), and of manganese bismuthide (Bismanol). These materials have higher coercive forces than the Alnicos, which renders them more suitable for certain applications, and moreover they are considerably lighter, which also increases the scope of their application. For example a minute rod of Ferroxdure magnetized at right-angles to its length is employed as the moving element in a recently designed gramophone pick-up.

There seems little doubt that these new materials will lead to fresh applications and to the further demand for even better magnets. How much further improvement is possible? If a material could be made up entirely of elongated particles of iron each small enough to be a single domain, and compacted so that its density is two-thirds that of the normal density of iron, it has been estimated that the coercive force would be about 10,000 oersteds, the remanent induction about 14,000 gauss, and the energy product about 40×10^6 gauss-oersteds. A glance at the table on page 179 shows that this is between five and ten times as great as that of the best materials now

* See Chapter 11.

The Development of the Permanent Magnet

available, and with this ideal material magnets would need to be only one fifth to one tenth of the size they are now. We do not yet know how to make such magnets, but there seems to be considerable incentive, which was not the case in the quest for the new high-permeability materials, to try and find ways in which this may be achieved, and it is likely that the future will bring not bigger, but smaller and better magnets.

Before going on to some of the applications of permanent magnets it is perhaps worth while adding a word or two about the care of permanent magnets. These may be supplied with a keeper – a small piece of ordinary iron which bridges the gap between the pole-pieces – and it is usually recommended that the keeper be placed on the magnet when not in actual use. In fact the keeper usually serves two distinct purposes. In the first place the magnet is usually magnetized with its keeper in place. Magnet and keeper then form a complete closed magnetic circuit. Consequently there is nowhere for poles to form, no demagnetizing field, and therefore a smaller magnetic field is able to magnetize it completely than would otherwise be the case. This fact is of some importance because it will be appreciated that a good permanent magnet material, being difficult to demagnetize, is also difficult to magnetize, and even in the absence of demagnetizing fields many thousands of amperes are needed to produce magnetic fields sufficiently great to magnetize it to saturation. When the magnetic field is removed the material is at remanence at A (Figure 38). When the keeper is removed the demagnetizing field causes the magnetization to drop to its operating value, namely that at point B. When the keeper is replaced the material should ideally go back to A. In practice there is some hysteresis and the material changes usually to a state of slightly lower remanence C. Removal and subsequent replacement of the keeper results in the magnetization tracing out the path CBC as shown.

The beneficial presence of a keeper results from two effects. Firstly with a keeper in place the magnet plus keeper form a closed ring which is almost immune to any demagnetizing fields which may exist due to other magnets nearby, or other unwanted influences. Secondly it will be noticed that at C the

Magnetism

magnetization curve (upper part) is more horizontal than at *B*. Now it is a fairly general rule that the effect of temperature variation and vibration on the magnetization of a substance is greatest where the slope of the magnetization curve (i.e. its inclination to the horizontal) is greatest. These influences do in fact have their maximum effect in the neighbourhood of the coercive force. Thus the magnetization of a magnet with its keeper in place is more stable with respect to these external

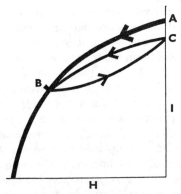

Figure 38. The action of a 'keeper' with a permanent magnet; for explanation see text.

agencies than it is without. However it must be stated that as a result of modern research present-day permanent magnet materials are quite remarkably stable even at their operating point and that with them the soft-iron keeper is no longer the essential item it was when the very best magnets were those made of carbon or tungsten steel.

The applications of permanent magnets are becoming so numerous that a complete list would not only increase the length of this book to encyclopaedic dimensions but would be out of date before it was printed. Instead we will content ourselves with a few cases which for one reason or another deserve special mention.

There appears to be an impression that for the purpose of scientific research the permanent magnet is obsolete and has

The Development of the Permanent Magnet

been replaced for all important purposes by the electromagnet. Twenty or thirty years ago this impression would have been more or less correct, but today the tendency is to replace even large electromagnets with permanent magnets wherever possible. There is everything to be gained from this, since, apart from the fact that a permanent magnet requires no external source of power to maintain its magnetic field, the field it provides is, with modern materials, far more constant than can be maintained with an electromagnet running off an ordinary electricity supply. Of course the field of a permanent magnet cannot be varied, except between experiments by re-magnetizing, but this is often not necessary for a large number of investigations. Permanent magnets have been extensively used in cosmic ray research and in the field known as radio-frequency spectroscopy which, unfortunately, is beyond the scope of this book. Another important use of permanent magnets is in electrical measuring instruments, galvanometers, ammeters, and voltmeters. The sensitivity of these instruments depends upon the strength of the magnetic field in which the coil is placed, and it is essential that the field be constant, otherwise the instruments' accuracy will not be maintained. Modern permanent magnet materials enable highly sensitive meters of this kind to be made which are extremely accurate, and which are robust and light enough to be easily portable.

A good deal of the research into new and improved permanent magnet materials was promoted by industries engaged in the manufacture of loudspeakers, which work on the same principle as that utilized in the galvanometer, namely that a wire carrying an electric current experiences a force in a magnetic field. The usual arrangement of the moving-coil loudspeaker is shown in Figure 39. The coil is wound on a light cylindrical former which is attached to a paper cone and suspended in the annular gap of a permanent magnet. As will be seen from Figure 39 the current in the coil is everywhere perpendicular to the magnetic lines of force, and so the force exerted on the coil is at right-angles to both these directions, into and away from the permanent magnet. The coil is therefore suspended in such a way that it can move in one direction only,

Magnetism

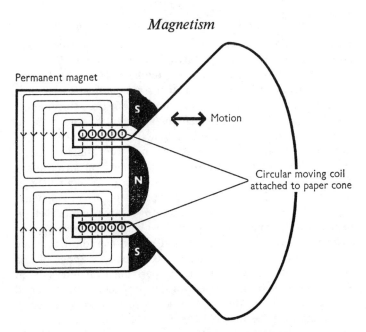

Figure 39. The permanent magnet moving-coil loudspeaker. The lines show the lines of induction within the magnet and the lines of magnetic field across the gap.

and when an alternating current from a radio receiver or gramophone amplifier is made to pass through this coil it moves in and out. This movement is transmitted to the cone diaphragm which gives rise to pressure variations in the air outside it, and thus acts as a source of sound. The force on the coil is proportional to the current flowing in it and to the strength of the magnetic field in which it is placed, and since the currents available to drive the loudspeaker are usually quite small it follows that the magnetic field must be as strong as possible if the loudspeaker is to work efficiently. In high-quality loudspeakers the magnetic field strength in the pole gap may be as high as 14,000 oersteds. Such loudspeakers respond equally well to a very large range of musical frequencies, and in consequence the sound which they emit bears a very faithful resemblance to the original.

The Development of the Permanent Magnet

A slightly different principle is used in the construction of the receiver which forms the essential part of the telephone earpiece. This consists of a thin flexible diaphragm made of a magnetically soft material, often silicon-iron, supported at its rim over the poles of a permanent magnet which has a coil of wire placed round it. The diaphragm, being magnetic, is magnetized by the permanent magnet and is in consequence attracted towards it. Suppose now a current passes through the coil. If its direction is such as to increase the magnetization of the permanent magnet, the pull on the diaphragm is increased and it moves nearer to it. If the current is in the opposite direction, on the other hand, it will recede. Thus if a current of suitable frequency is passed through the coil the diaphragm will move in and out and will generate a sound in the air just in the same way as a moving-coil loudspeaker. This type of receiver is much more sensitive than the latter and has to be used in the telephone earpiece on account of the minute currents available from ordinary telephone lines. The choice of magnetic material is important. It is necessary to have a permanent magnet otherwise the diaphragm would be pulled towards the coil no matter what direction the current flowed in it, but it is essential that its magnetization be changed somewhat by the current in the coil. In technical terms this means that the material must have a fairly high permeability (slope of magnetization curve) at its operating point, and this requirement is met fairly well by certain cobalt-steel magnets. This type of receiver, although sensitive, does not respond very well to high or low frequencies, and for this and other reasons the sound heard through a telephone earpiece is nothing like as faithful a reproduction of the original as can be obtained from a moving-coil loudspeaker.

As is well known the attraction between a permanent magnet and a piece of iron is quite large, and with a really strong permanent magnet a very large force is necessary to separate them when they are in contact. This fact is used in the construction of the magnetic chuck – a device for holding articles of iron and steel in place while they are being machined, which avoids the use of mechanical clamps and screws which may either harm

Magnetism

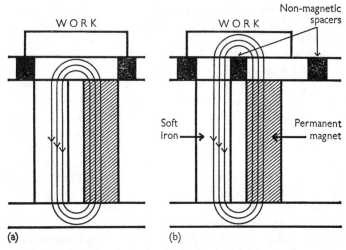

Figure 40. The magnetic chuck; for explanation see text.

the work or may otherwise be impractical. For example, with thin discs whose faces require grinding it is very difficult to find a suitable means of fixing them in position for the operation. The principle of the magnetic chuck is shown in Figure 40. The essential features are the solid vertical permanent magnet blocks, the base plate of soft iron, and the movable top plate which consists of magnetic (soft iron) and non-magnetic segments. When the top plate is in position (a) the magnetic lines of force from the magnet are returned through the upper and lower plates which form a closed magnetic ring with no poles and which consequently exert no force externally. When the top plate is made to slide sideways the non-magnetic segment cannot provide a path for the lines of force, which have to pass through the work to be held, and over the other side of the segment, in order to return to the magnet. In this position there is a very strong force acting on the work, sufficiently strong to secure it in position while it is being machined.

Further examples of the use of permanent magnets are to be found in small electric motor-car speedometers, magnetic electric clocks, electricity meters, as focussing devices in tele-

The Development of the Permanent Magnet

vision receivers, magnetic door catches, and even in children's toys. The list could be extended if any useful purpose could be served thereby. Instead, our final example is of a recent development which has made great advances in recent years and which is slightly different in principle from those so far described – magnetic tape recording.

Although it is only in recent years that magnetic recording has started to come into general use, its invention by a Danish engineer V. Poulsen dates back to 1898. He constructed an apparatus in which a steel wire wound on a cylinder was moved between the two pole pieces of a small electromagnet which served for recording as well as reproducing. In this way he was able to record the current from a microphone on the steel wire and by moving the wire in the opposite direction he was able to make the record audible with an earphone. His apparatus, called the Telegraphone, was demonstrated at the Paris exhibition of 1900. By this time however the first commercial gramophone record had just been made. It must be remembered that no apparatus was at that time available for amplifying small electrical signals (indeed this did not become available until nearly twenty-five years later), and so interest in the Telegraphone was not maintained, and instead further research was devoted to the gramophone record and the reproduction of sound by purely mechanical means.

The basic idea of magnetic recording is to convert sound waves into electrical vibrations and to send these electrical vibrations through a small electromagnet which induces magnetization in a moving steel wire, which is proportional to the strength of the electric current (and hence to the loudness of the original sound) at any instant. By 1925 amplification of tiny electrical currents had become perfectly feasible, and the limitations imposed on the success of magnetic recording were those of the magnetic characteristics of the moving wire rather than in the recording and reproducing system. For if an ordinary magnetic material is magnetized locally, the magnetization tends to spread somewhat at the expense of the small area originally magnetized. What is really needed is a material which retains its magnetization but which is itself difficult to

Magnetism

magnetize, so that a small area may be magnetized without affecting the material surrounding it. Now these are the characteristics of a good permanent magnet material. Unfortunately until recently all the suitable permanent magnet materials were brittle and for this reason unusable in tape or wire form.

Towards the end of World War II British observers listening to German radio broadcasts noticed that certain recorded programmes were being transmitted with a fidelity which could not at the time have been obtained from the normal gramophone discs. The absence of breaks in the transmission suggested that some form of tape or wire recording was being used, and it soon became clear that the problem of finding a suitable magnetic material had been solved. The material was a magnetic oxide of iron known as $\gamma\text{-}Fe_2O_3$* in finely powdered form thinly coated on celluloid tape. It is now fairly certain that the particles of this material, which are formed chemically in the form of thin needles, are of a size sufficiently small for them to be single domain particles. When mixed together with a suitable glue and painted on thin celluloid the resulting tape has a coercive force of about 250 oersteds and the individual particles, being single domains, do not interfere with each other very much. Consequently it is possible to have adjacent areas separated by only a fraction of millimetre magnetized in opposite directions, with the result that sound waves of very high frequencies can be recorded without excessive tape speeds. The way in which the tape is magnetized is shown in Figure 41. The essential feature is a small electromagnet containing a core of a high-permeability material the ends of which are separated by a minute distance so that the field in the gap is very intense and acts only over a very short distance. In 'playing back', this high-permeability core becomes magnetized by the moving magnetic tape and induces an electromotive force in the surrounding coil which can be amplified and made to operate a

* Not to be confused with $\alpha\text{-}Fe_2O_3$, the material haematite. The prefixes α and γ indicate the different types of crystal structure of the same compound. Haematite is very weakly magnetic – ostensibly paramagnetic but actually much more complex in its behaviour. $\gamma\text{-}Fe_2O_3$ has the same crystal structure as Fe_3O_4 (the mineral lodestone) and like it is ferromagnetic (but see Chapter 11) but with a smaller spontaneous magnetization.

The Development of the Permanent Magnet

loudspeaker. The modern tape recorder provides a good example of the synthesis of the use of both soft and hard magnetic material, for without either its performance would be so seriously impaired as to offer very little advantage over the

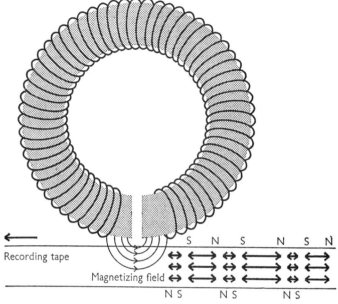

Figure 41. The principle of the magnetic tape recorder.

conventional method of recording on disc. Instead its use has become standard for all forms of serious recording and in common with all other purely technical advances has already served as a useful investigational tool in scientific research.

CHAPTER 11

Ferrimagnetism and Antiferromagnetism

IN addition to the ferromagnetic metals and their alloys there exists a group of substances which exhibit strong ferromagnetic properties but which cannot strictly be said to belong to the same class. For one thing they are not metals, but oxides of iron mixed with other metal oxides. For another their saturation magnetization is invariably less than that of nickel. The best known example is the magnetic oxide of iron whose chemical formula is Fe_3O_4 – lodestone. The existence of chemical compounds of iron oxide with other oxides has been known at least since 1860, and their ferromagnetic properties were known by 1887. These compounds remained curiosities until about 1930, and the reason for this no doubt lay in the fact that there is such a bewildering variety of possible compounds that no one contemplating a serious study of their properties could see where to begin. By this time however the use of X-rays was becoming widely adopted as a means of determining the crystal structure of a chemical compound and the positions its constituent atoms occupied within it.

There was however another reason for investigating their properties – the possibility that they might be of some use. Ordinary ferromagnetic materials, both hard and soft, although admirable in most respects do possess certain disadvantages for certain applications. For one thing they are heavy, and it is desirable in modern aircraft, for example, with their utter dependence upon electrical instruments for navigational and control purposes, that they do not carry a bigger permanent load than is absolutely necessary. Nevertheless it is doubtful whether this by itself would have provided the necessary stimulus for further research. This came from the electrical industries, most of all from those involved in radio transmission and reception and in the general field of telecommunications. To see why this was so consider the problem of sound

Faraday's electromagnet with which he first discovered diamagnetism. The yoke of the magnet was constructed from a link of a great iron chain.

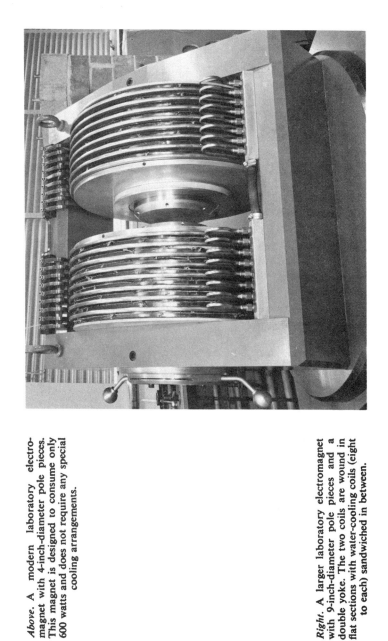

Above. A modern laboratory electromagnet with 4-inch-diameter pole pieces. This magnet is designed to consume only 600 watts and does not require any special cooling arrangements.

Right. A larger laboratory electromagnet with 9-inch-diameter pole pieces and a double yoke. The two coils are wound in flat sections with water-cooling coils (eight to each) sandwiched in between.

Above. The world's largest laboratory electromagnet, at Bellevue, Paris. This magnet weighs 120 tons. It has a large number of interchangeable pole pieces permitting wide variations of field conditions.

Right. The electromagnet of the Bevatron (synchrotron) at the University of California. This magnet, which weighs 10,000 tons, is 14 feet high and over 130 feet in diameter. The field strength in the gap is about 16,000 oersteds.

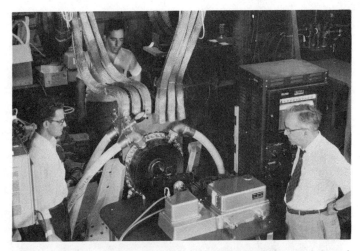

(a) Bitter-type solenoid, here being used in an experiment on infra-red Faraday rotation. The coil consumes 10,000 amperes (note the flexible metal tape used to lead the current in and out) at 170 volts. Water is circulated through four 2-inch fire hoses at 800 gallons a minute to provide the necessary cooling. Professor Bitter on the right.

(b) The generator at the Massachusetts Institute of Technology used to provide the power for the Bitter solenoids. It provides 170 volts at 10,000 amps. Note that the entire output from this generator is being put through the solenoid shown in (a). Standing: Dr H. H. Kolm and Professor F. Bitter.

Flat spiral used in constructing Bitter solenoids. Water is forced through the holes under pressure.

(*a*) 1,500 kilowatt solenoid designed and constructed at the Clarendon Laboratory, Oxford. It takes a maximum current of 4,500 amps and produces a field of 55,000 oersteds. It consists of four flat coils each wound with copper strip somewhat like a very long clock-spring contained within a fibreglass case.

(*b*) This shows the construction of the coils. The white objects are the remains of ceramic spacers after a slight mechanical failure.

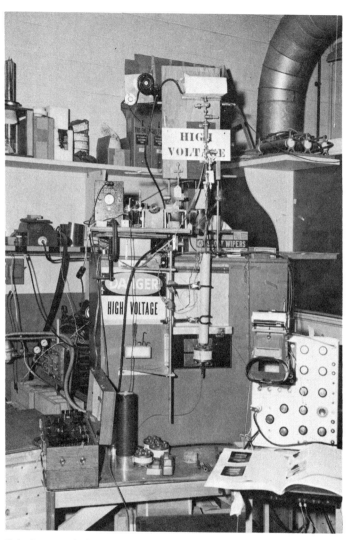

Pulsed magnetic fields. This photograph shows an experimental arrangement using pulsed magnetic fields. The solenoid is seen in front of the middle of the large box in the centre which contains the condensers (ten at 200 microfarads, each charged to 3,000 volts). A liquid helium dewar protrudes from its lower end. This solenoid produces fields of about 400,000 oersteds for about one thousandth of a second. Three other solenoids can be seen on the bench in the foreground.

This photograph shows the details of construction of the solenoid shown in the previous photograph. The coil is machined out of an alloy of beryllium and copper and compressed between two brass end-plates which serve as current leads. A solenoid of this type with a diameter of $\frac{3}{16}$ inch will produce a field of 750,000 oersteds.

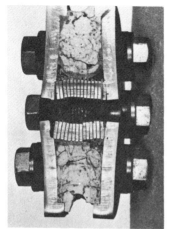

Four solenoids of the type shown in the previous two photographs, cut open after use. The uppermost show mechanical distortions due to electromagnetic forces. The lower left is a 750,000 oersted solenoid with a ⅜ inch internal diameter after much use, and the lower right is a rather larger coil producing a smaller field. Both these coils are adequately supported to withstand electromagnetic forces.

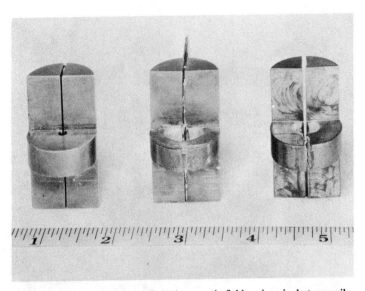

PLATES 28 AND 29: *Above.* Pulsed magnetic fields using single-turn coils. By this means magnetic fields of 1.6 million oersteds have been obtained. The plate above shows the coil before (left) and after (centre and right) use.

Right. This shows the state of the metal after the coil has produced a field of over a million oersteds. The distortion is caused by the combined effect of momentary melting (clearly visible in *c*) and the intense electromagnetic forces which produce a pressure outwards tending to expand the coil.

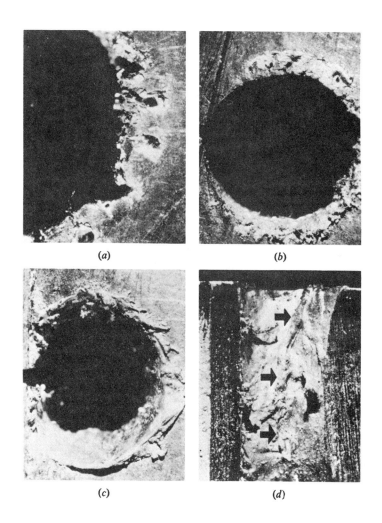

Magnetic fields as a research tool. This photograph shows a solenoid somewhat similar to that of Plate 24 being used in a nuclear demagnetization experiment at Oxford. With this magnet the lowest temperature yet achieved (about $15 \times 10^{-6}\,°K$) was obtained.

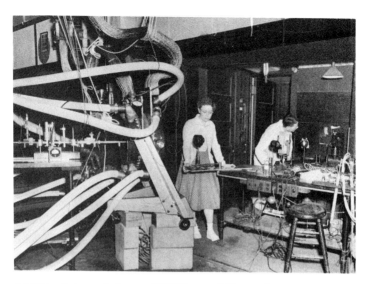

(a) This photograph shows a 105,000 oersted Bitter solenoid mounted for use in an optical Zeeman effect experiment.

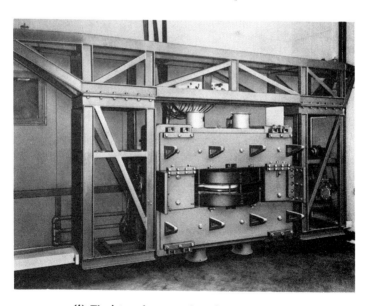

(b) The internal construction of a large betatron.

A magnetron complete with permanent magnet. The fins which can be seen emerging from the pole gap are to provide air-cooling.

Ferrimagnetism and Antiferromagnetism

broadcasting. In the nineteen-twenties the most powerful transmitters were those emitting long waves – the long-wave transmitter at Droitwich operates at 1,500 metres at a frequency of 200,000 cycles per second. This is the so-called carrier frequency. To transmit sound waves these are first converted into electrical oscillations and then superimposed on the carrier wave. To transmit sound faithfully requires frequencies up to at least 10,000 cycles a second, and when these waves are added to the carrier its frequency varies between 200,000 plus 10,000 and 200,000 minus 10,000, i.e. between 190,000 and 210,000 cycles per second. If another transmitter operates at a carrier frequency of 210,000 cycles per second this will be modified to a range of frequencies between 200,000 and 220,000 when transmitting sound, and any receiver situated midway between them will be unable to separate one from the other. Now this is exactly the situation on the European medium-wave band today, as most people will have realized when trying to listen to a broadcast from a distant transmitter. One way of avoiding this overcrowding is to transmit at a higher carrier frequency. Most radio receivers can be tuned over a range of carrier frequencies of one octave, that is from one frequency to double that frequency. Now from 100,000 to 200,000 cycles per second, allowing the transmitter frequencies to be spaced by 25,000 cycles to avoid serious interference, there is room for only four stations. From 1 to 2 megacycles per second (1 megacycle = 1 million cycles) there will be room for forty, and from 10 to 20 megacycles per second, four hundred. Short-wave transmission has other advantages over long-wave, but the essential point which should be emphasized is that the more widespread radio-communication becomes the more it becomes necessary to employ short waves and high frequencies. The same is true of carrier telephony and, to a certain extent, of television broadcasting.

Unfortunately high frequencies are attended by their own problems, not the least of which is that ordinary ferromagnetic substances are of little use. This is because of eddy-currents, previously mentioned in connexion with transformer materials. These currents are always induced by alternating magnetic

Magnetism

fields and their effect is to confine the magnetization closely to the surface of the material, leaving the interior virtually unaffected by the field and therefore unused. This can be avoided to a certain extent by using the material in the form of thin sheet. Unfortunately it has been found necessary to reduce the thickness of ordinary transformer sheet to the thickness of only fourteen thousandths of an inch when operating at the normal mains frequency of 50 cycles a second. To be equally efficient at a frequency of 50 megacycles a second (a typical television frequency) the thickness would have to be reduced to fourteen millionths of an inch, and although metals have been rolled to thicknesses of as little as one hundred thousandth of an inch, to go very much further than this appears to be a practical impossibility. It is evident that ordinary ferromagnetic material cannot be used with any advantage at these frequencies.

Yet there is no doubt that if eddy currents and their effects could be eliminated magnetic materials would be just as useful at high frequencies, in improving the efficiency of transformers and reducing the size of coils, as they are at low frequencies. Consequently various attempts were made to produce materials in which eddy-currents were not present or at least kept to a minimum. One method which was very successful was to use finely powdered iron cemented together with an electrically insulating glue to form a solid block. It is of course very necessary to ensure that the particles of iron are not too small, otherwise they will behave like single domains and the material will have a low permeability and all the properties of a permanent magnet. With care this can be avoided however, and the resulting material can be used satisfactorily up to about 50 Mc/s frequencies.

Meanwhile it had occurred to J. L. Snoek, at the Philips Laboratories in Holland, that the iron-metal oxides, now known collectively as ferrites, might repay further study, for they possess the unusual feature of being electrical insulators, intrinsically and not merely by virtue of being mixed together with insulating cement. Eddy currents and their effects are therefore absent and they can in principle be used effectively at

Ferrimagnetism and Antiferromagnetism

all frequencies no matter how high. The problems facing Snoek were first to find out exactly what these ferrites were, for even the exact composition was unknown at the time, why they are magnetic, and whether they could ever be manufactured to specific and reproducible standards. Broadly speaking he succeeded in the first and third of these, but the elucidation of their magnetic properties was first carried out by L. Néel shortly after World War II.

Ferrites are definite chemical compounds having the general chemical formula $MO.Fe_2O_3$, in which M stands for any divalent ion, for example copper, zinc, cadmium, magnesium, or even iron itself. In the case of the latter we get $FeO.Fe_2O_3$, which is another way of writing Fe_3O_4, the chemical formula for magnetite or lodestone. Mixed ferrites are also possible, for example $(\frac{1}{2}Cu.\frac{1}{2}Zn)O.Fe_2O_3$, and in fact a whole range of intermediate compounds from $CuO.Fe_2O_3$ to $ZnO.Fe_2O_3$ are possible, thus indicating that ferrites can form alloys just like metals. Not all the ferrites are ferromagnetic but many are, and it is this, coupled with their ability to form alloys, which makes such a vast range of magnetic properties possible. Ferrites are made by mixing together the separate oxides in the correct proportion and then heating the mixture to a high temperature, usually between 1200° and 1350° C for several hours. This process is known technically as sintering. A definite chemical composition is established and the resulting substance is chemically homogeneous although it may be porous and mechanically weak. Nowadays ferrites can be produced which are almost completely free from porosity and extremely hard. They are invariably black and can often take a high polish.

Investigations using X-rays show that they crystallize in the so-called spinel structure, named after the mineral spinel, which has the chemical formula $MgO.Al_2O_3$, of the same form as that of the ferrites. This is shown in Figure 42, and if at first sight it appears unusually forbidding we may take consolation in the fact that the structure is essentially cubic like that of iron and nickel. Ferrites are electrical insulators for the same reason that sodium chloride, common rock-salt, is also an insulator. The spheres in Figure 42 represent metal, iron, and oxygen

Magnetism

ions. Ferrites are in fact basically ionic compounds, just like rock-salt, and are held together by the electrical forces of attraction between the positive charges on the metal ions and the negative charges on the oxygen ions. In the process of ion formation electrons are transferred from metal to oxygen, and if the number of oxygen atoms is just right every electron given up by the metal atoms will be received by the oxygen atoms. None are left over to become free electrons as in a metal, and the substance is an insulator. It is very important to get the right amount of oxygen, however, for if there is too

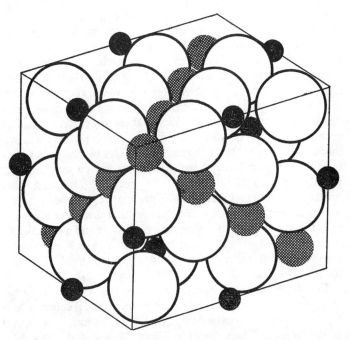

Figure 42. The spinel structure. The large white spheres represent oxygen ions; the black spheres surrounded by four oxygen ions are known as A-sites; in a normal spinel these are occupied by the divalent ions. The shaded spheres surrounded by six oxygen ions are known as B-sites; in a normal spinel these are occupied by trivalent ions (the Fe^{+++} ions in 'normal' ferrites).

Ferrimagnetism and Antiferromagnetism

little there may be a slight excess of electrons, while if there is too much there will be a small deficit, both of which will lead to electrical conductivity. This necessity for exactness of composition is by no means confined to ferrites, for it is a general feature of a large group of materials known as semi-conductors. Ferrites are in fact rightly classed as semi-conductors, among whose chief characteristics is that any deviation from the exact chemical formula inevitably decreases the electrical resistance, and that the resistance of these materials decreases rapidly with increasing temperatures (unlike metals, whose resistance increases slightly in similar circumstances). Keeping the composition exactly right during the sintering process is in fact one of the secrets of the manufacture of ferrites, for although the magnetic properties are not influenced by departures from chemical exactness, the electrical properties are, and it is primarily for their electrical properties that they were originally developed.

The fact that these substances are ionic should make their magnetic properties very simple to calculate, for we know from investigation on paramagnetic salts what the magnetic moment of the various metal ions should be, so that if, as in ferromagnetic substances, all the metal irons are aligned parallel by an 'internal field' then the observed spontaneous magnetization or saturation magnetization should be equal to the sum of the magnetic moments of the individual ions. Take as an example the case of magnetite. It will be noticed in Figure 42 that there are two essentially different positions in which a metal ion may fit between the oxygen ions, and these have been drawn as black and shaded spheres. The black spheres are surrounded by four oxygen ions and are known as A sites; the shaded ones are situated between six oxygen ions and are known as B sites. In a normal spinel the divalent metal ions occupy the A sites and the iron ions the B sites. Equally often the trivalent (iron) ions occupy A sites, the B sites being occupied by the divalent ions and the remaining half of the trivalent ions. Such spinels are known as inverse. Magnetite has the structure of an inverse spinel and consequently each molecule contains one Fe^{+++} ion on an A site and one Fe^{++} ion and one Fe^{+++} ion on a B site.

Magnetism

The magnetic moment of the Fe^{++} ion is known to be 4 Bohr magnetons, that of the Fe^{+++} ion 5 Bohr magnetons, and thus we should expect each molecule to have a magnetic moment equal to $5 + 5 + 4 = 14$ Bohr magnetons. The observed value is 4·08. Attempts to calculate the saturation magnetic moments of all other ferrites are no more successful, the observed values always being much less than those calculated.

Now we do not necessarily expect our estimates to agree exactly with the values observed by experiment, but neither do we expect them to be so seriously in error as in this case. When this is so it is likely that our basic hypothesis is unsound, and in 1948 Néel suggested that the assumption of parallel alignment of the magnetic moment of the individual ions might not be true in ferrites, and he showed how an alternative arrangement might be achieved. Suppose the ions on the A sites in magnetites have their magnetic moments aligned parallel to each other and those on the B sites are also aligned parallel to each other but in the opposite direction. The magnetic moments of the ions on the two types of site will partially neutralize each other and the resultant magnetic moment will be $4 + 5 = 9$ Bohr magnetons on the B sites minus 5 Bohr magnetons due to the Fe^{+++} ions on the A sites, thus 4 Bohr magnetons altogether, which is very close to the observed value. Similar calculations based on this hypothesis predict the saturation magnetization of all other ferrites with considerable accuracy, thereby lending considerable support to the view that it is essentially correct.

Now is this hypothesis of Néel a reasonable one? If we recall to mind the early part of Chapter 7 we will remember that ferromagnetism occurs because of a certain interaction between the magnetic moments of the atoms which causes their spin moments to become aligned parallel with each other. This interaction is termed exchange, and the tendency towards parallel alignment of the spins is the consequence of a certain mathematical quantity known as the exchange integral having a positive sign. In the hydrogen molecule the exchange integral is negative and this favours alignments of the two electron spins in opposite directions so as to oppose each other. Now, as stated before, this exchange integral depends, amongst other

Ferrimagnetism and Antiferromagnetism

things, on the distance between neighbouring atoms, and it so happens that in ferrites the distance between the ions on both the A and B sites is such as to make the exchange interaction between them positive and therefore to promote parallel alignment of the spins. At the same time the distance between the ions on the A sites and those on the B sites is such as to make any exchange interaction between them negative and thereby to favour a situation in which the spins of all the ions on the A sites point in the opposite direction to all those on the B sites. Thus Néel's hypothesis is perfectly reasonable in the light of our knowledge of the interactions which exist between atoms or between ions.

Néel coined the term ferrimagnetism to cover the magnetism of ferrites and worked out the consequence of his hypothesis in a remarkably complete manner. Each set of ions behaves like an ordinary ferromagnetic substance with a characteristic Curie temperature above which the parallel alignment is destroyed by temperature. It is not necessary that the Curie temperature of the A site and B site ions be the same, and in consequence, although the spontaneous magnetization of a ferrite disappears abruptly at some definite temperature, as is the case with ferromagnetics, the manner in which its magnetization varies with temperature is not the same. Indeed, in a ferrite in which the A site ions have large magnetic moment but low Curie point and the B site ions small magnetic moment (oppositely directed) and high Curie point, it is possible for the net magnetization of the ferrite to change sign as the temperature is raised. Certain ferrites have been found in which this is actually observed and this constitutes very powerful evidence for the existence of negative interaction and antiparallel alignment of spins. The temperature at which the magnetization of a ferrite finally becomes zero is usually known as the Néel temperature.

We can now understand why the spontaneous magnetization of the ferrites never approaches that of iron, for although the magnetic moment of the individual ions may be quite large they always cancel each other out to a certain extent, and the magnetization which is left is diluted by the presence of the non-magnetic oxygen ions. This has led to attempts to make ferrites containing ions of the rare earth metals, which have larger magnetic moments than those of the iron group of metals. Some of these have been successful, notably the rare earth garnets of composition $R_3Fe_5O_{12}$, where R is a rare earth ion or an yttrium ion.

In most respects ferrites have magnetic properties which are very similar to those of ordinary ferromagnetic materials. They exhibit the same phenomena of hysteresis and magnetostriction. Moreover single crystals of ferrites show precisely the same kind of anisotropy as that observed in iron or nickel. Thus the same guiding principles may be used to make magnetically soft and hard ferrite materials as are used with the more usual ferromagnetic metals, the only difference being that there are more different ferrites to start from. In particular, soft ferrites having a high initial susceptibility are those in which the magnetostriction is small, and it just happens that whereas the magnetostriction of ferrous ferrite $FeO.Fe_2O_3$ (magnetite) is positive, that of all the others is negative. Consequently by mixing any ferrite with magnetite in suitable proportions a mixed system may be made whose magnetostriction is zero and which therefore has a high initial susceptibility.

Some of these ferrites have certain special properties which has brought them into widespread use. Magnesium ferrite for example has a low coercive force and a rectangular hysteresis loop which appears to be an intrinsic property not requiring any special mechanical or heat treatment to bring it about. Consequently it is an ideal material for cores which make up the memory units in computers. Cobalt ferrite has a very large magnetostriction – about twenty times that of nickel at saturation, and this makes it a useful material for producing ultrasonic vibrations. Only the fact that nickel is still mechanically the stronger of the two materials has prevented it from being replaced altogether. Certain other ferrites, notably barium and cobalt ferrites, have high coercive forces, particularly in powdered form, and from these good permanent magnets, known under the trade names of Ferroxdure and Vectolite respectively, have already been commercially produced. Ferroxdure has a

Ferrimagnetism and Antiferromagnetism

coercive force of about 2,000 oersteds and an energy product of about 3×10^6, and can therefore be made into short squat magnets of great strength. This is very useful when space is limited and together with the fact that a Ferroxdure magnet weighs only about half that of an Alnico magnet of the same size, promises it a great future.

It is however for operation at high frequencies that ferrites are particularly suited, for they are electrical insulators and do

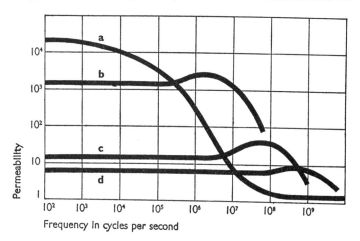

Figure 43. The variation of the initial permeability of various materials with frequency. Note the compression of both vertical and horizontal scales. (a) Typical high-permeability nickel-iron alloy one thousandth of an inch thick. (b) Manganese-zinc ferrite. (c) A nickel ferrite known commercially as Ferroxcube. (d) A typical ferrite of the 'Ferroxplana' class. Note the superiority of this over all other materials at frequencies greater than 10^8c/s.

not suffer from the effect of eddy-currents. They can therefore be used as cores in coils and transformers in radio and television receivers at frequencies at which ordinary ferromagnetic material would be no more than an encumbrance. Figure 43 shows for comparison the way the effective permeability of typical substances varies with frequency, from which the superiority of ferrites at high frequencies may be seen at a

Magnetism

glance. The question may be asked why the effective permeability depends upon frequency at all. With metals it is easy to understand because the eddy-currents confine the magnetization to a thin layer near the surface and with increasing frequency the eddy-currents become greater and the layer thinner until eventually only a minute fraction of the metal is being magnetized at all. Very little difference would be made by removing it altogether and so its effective permeability is not significantly different from that of air. With ferrites it is much less simple, for there are no eddy-currents and what causes the reduction in their permeability is still something of a mystery. Certain causes are known, and the development of Ferroxplana (Figure 43) was the outcome of a deliberate and reasoned attempt to overcome them, and not, as might be supposed in view of the fact that ferrite development is still in its infancy, a happy accident arising from the chance mixture of two ferrites by a physicist or chemist with magnetically 'green' fingers.

It may happen that the magnetic moment of the ions on the A sites is exactly the same as the combined magnetic moments of those on B sites. The resultant magnetic moment of the system will be zero in this case, and in spite of the internal alignment such a substance will not show any ferromagnetic properties but will behave to a certain extent like a paramagnetic substance. Such substances are known and are described as antiferromagnetic. Antiferromagnetism is not confined to those materials having the spinel type structure or solely to oxides. Some of the best known are copper chloride, $CuCl_2$, and the oxides of manganese, cobalt, and nickel; certain metal alloys are also known to be antiferromagnetic. They have a small positive susceptibility and their magnetic behaviour is different from that of a typical paramagnetic only in that, instead of obeying a Curie or Curie-Weiss Law, their susceptibility increases with decreasing temperature, but then decreases below a certain critical temperature. The temperature at which the susceptibility is a maximum is known as the Néel temperature, and it is here that spontaneous antiparallel alignment of the magnetic moments sets in. The Néel temperature in antiferromagnetics is thus exactly analogous to the Curie

Ferrimagnetism and Antiferromagnetism

temperature in ferromagnetics, the only difference being that whereas in the latter the spins are aligned parallel to each other below the critical temperature, in antiferromagnetics the spins are anti-parallel.

We can readily appreciate that the tendency towards antiparallel alignment of the magnetic moments of the ions or atoms from the same type of exchange forces as those responsible for ferromagnetism but with the difference that the exchange integral is negative (as in the hydrogen molecule). This can be regarded as being equivalent to an internal field which always tries to decrease the magnetization. Thus we might expect that instead of obeying a Curie-Weiss Law the susceptibility of an antiferromagnetic substance above its Néel temperature should obey a law of the form:

$$\kappa = \frac{C}{T + \theta}$$

with a positive sign instead of a negative one in the denominator, and such substances are indeed observed to behave in accordance with an equation of this type in most cases.

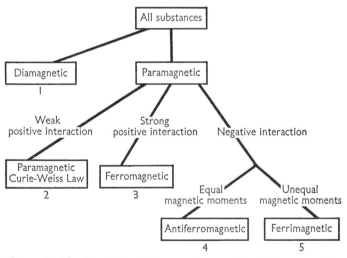

Figure 44. The five different classes of magnetic substance and the way in which they arise.

Magnetism

At present antiferromagnetic substances have no application of a scientific or practical nature and are to a certain extent the 'Cinderellas' of magnetism. Their importance to the physicist lies more in the fact that they are a logical necessity whose existence is required to complete the full range of magnetic substances made possible by the presence of positive and negative interaction between atoms or ions. The connexion between the various types of magnetic substance is shown schematically in Figure 44, from which it will be seen that we now recognize five different classes. However, Faraday's original classification into three types according to their gross magnetic properties is still sufficient for most purposes. It is only from an atomic standpoint that a further sub-division becomes necessary.

CHAPTER 12

Magnetism in Scientific Research

IT is probably true to say that of all the experiments carried out during the past sixty years which have led to major scientific advances in physics something like half have depended for their operation on the use of magnetism. Furthermore we have seen that magnetism is in fact a universal property stemming from the very heart of a substance, from its electrons and the way they are arranged, and one of the standard ways of obtaining information about the arrangement of electrons within an atom of a substance is to place it in a magnetic field and by various means to observe its behaviour. We cannot do this until we have first produced the magnetic field into which the substance is to be introduced, and in this type of experiment and almost all others the primary requirement is a magnetic field as strong and extending over as large a space as possible. Before discussing the part played by magnetic fields in modern physical research we shall do well first of all to pay some attention to the way they are achieved.

A magnetic field can be produced either by a permanent magnet or an electromagnet, and the present tendency is to use permanent magnets whenever possible. They have two serious drawbacks however, namely that the magnetic field cannot be varied continuously and that it is usually not possible to obtain from them fields greater than about 10,000 oersteds. If these restrictions are unimportant there is everything to be gained by using a permanent magnet. Otherwise recourse must be made to an electromagnet. These have the great advantage of flexibility, for their magnetic field may be varied quite simply by altering the electric current flowing in the coils, and moreover the field which they can provide is limited only by the current which is available for their operation.* It is possible to

* This statement is not true in principle; the ultimate limit is set by the mechanical strength of the coils; in practice the maximum field which can be attained is usually limited by the power available for supplying the magnet.

Magnetism

divide electromagnets into two classes – those containing an iron core and those without. The simplest type of electromagnet is the solenoid – a long coil of wire in which the magnetic field is directly proportional to the number of turns of wire divided by the length of the coil, and to the current flowing in it. Simple solenoids wound with many turns of wire are very convenient for the production of small magnetic fields, but they have one great drawback which limits the attainment of fields greater than about 1,000 oersteds except for very short periods of time. This is that the electric current generates heat in the windings in overcoming the electrical resistance of the coil, and unless some provision is made for removing this heat the coil becomes hot. The first result of this is that the current begins to decrease, because the current is inversely proportional to the resistance of the wire and this increases as the temperature rises. This is in itself undesirable if a steady magnetic field is required, as is usually the case, but what is more serious is that the heating, if allowed to continue, may become so severe that the insulation on the wires becomes burnt and the solenoid is ruined.

There are two ways of overcoming these difficulties, one of which is to make adequate provision for the removal of the heat generated, the other being to concentrate the magnetic field using an iron core. Until fairly recently the latter procedure was invariably adopted, so much so that the word electromagnet conjures up in most people's minds a picture of iron pole pieces with massive coils of wire round them. To them it may come as a surprise to learn that this form of electromagnet is limited to the production of magnetic fields less than about 50,000 oersteds. For fields greater than this the presence of the iron core ceases to be sufficiently beneficial to warrant its presence and the solenoid comes back into its own. For what we might term intermediate fields and when not too large a field is required to act over a very large space, the iron-cored electromagnet has certain definite advantages which have made it a universal piece of equipment in most research laboratories.

The essential features of the iron-cored electromagnet may be seen from Plate 18. They are: (1) the so called yoke consisting

Magnetism in Scientific Research

of a U-shaped piece of soft iron having a rectangular section, (2) the pole pieces, also of soft iron and usually circular in cross section, and (3) the two magnetizing coils. The use of soft iron is to ensure that the magnet is easily magnetized and exhibits little hysteresis, for we do not want its magnetization and therefore the strength of the magnetic field to depend on the direction in which the current was previously applied. When a current flows in the coils, the iron becomes magnetized and a field is produced in the gap between the poles just as with a permanent magnet. The strength of the field in the gap between cylindrical pole pieces should be equal to the magnetic induction in them. This is equal to $H + 4\pi I$ and for iron $4\pi I$ is about 20,000, so that any field in excess of this amount must of necessity come from H, the field due to the coils alone. In practice the field is always less than this, because even with cylindrical pole pieces the lines of force do not go straight from one pole face to another but always curve outside slightly, and this so-called 'leakage' reduces the field in the gap. Some increase can be obtained by tapering the pole pieces, but this is very small unless the reduction in the cross-section of the pole pieces is made very great. The increase in the field obtainable by this means is of course accompanied by a decrease in the volume over which it acts, so that the effect of tapering is to concentrate rather than increase the field. One way of genuinely increasing the field in the gap is to use pole pieces which are tipped with an alloy of iron and cobalt. This has a somewhat higher saturation magnetization than pure iron and consequently the field in the gap is correspondingly greater. The only other way of increasing the field is to increase the current in the coils so that their magnetic field adds appreciably to that of the iron. This is a perfectly legitimate procedure of course, but it ultimately brings us back to the solenoid and to the problems of removing large quantities of heat from the windings. The real advantage of the iron core, that of being easily magnetized and therefore producing quite large fields with low currents and the expenditure of only small amounts of electrical energy, thereby becomes lost.

We can illustrate this point quite well by drawing attention to the serious difficulties that arise when the size of this type of

Magnetism

magnet and the field it is required to give are increased. Plate 18 shows a medium-sized electromagnet used for general research purposes. Its pole pieces are four inches in diameter and the whole magnet weighs about $5\frac{1}{2}$ cwt. It may be run off 200 volts D.C. mains supply and will produce a field of about 8000 oersteds over a volume of a few cubic centimetres with a current of 8 amps. This involves an expenditure of power of only 600 watts, and the magnet can be run continuously at this rate for several hours without overheating. No cooling arrangements are thus necessary. If larger magnetic fields than this are required continuously then arrangements have to be made for getting rid of the heat and larger magnets than this almost invariably require some form of cooling. One of the simplest ways of doing this is to pump oil (which is a good electrical insulator) through the gap between the windings and to cool the oil by passing it through tubes immersed in water. Unfortunately this is not a very efficient way of cooling, and when large amounts of power are being dissipated the oil cannot carry the heat away quickly enough. The next step is to wind the magnetizing coil, not with solid wire, but with copper tubing and to remove the heat by passing cooling water through it. This principle is used in many of the large magnets used for general research purposes. An example of a larger electromagnet is shown in Plate 19. As the effect of the iron is to concentrate rather than increase the available field, this latter depends very much on the separation of the pole pieces and on the extent to which they are tapered. Consequently provision is made for changing them so that with a little rearrangement we can obtain either a large field in a small volume or a more moderate field over a large volume and so on.

There are a number of really large general-purpose electromagnets in various laboratories all over the world, the largest being that at the Académie des Sciences in Paris. This magnet, which weighs 110 tons, is shown in Plate 20. Its great advantage over smaller magnets is that quite large fields can be maintained throughout large volumes. For instance, a field of 40,000 oersteds may be obtained throughout a volume of about 20 c.c. (only about a one and a half inch cube, but this would be

Magnetism in Scientific Research

regarded as extremely large for certain measurements). In order to achieve this something like 100 kilowatts of power are required. With the expenditure of 1,000 kilowatts* a field of 100,000 oersteds can be obtained over a volume of about one cubic centimetre. Of this probably not more than half comes from the iron core, and so the stage has been reached at which the utility of the iron is decreasing. In order to double this field over an appreciable volume the coils would have to supply three times as much field as the iron, and at this stage it is better to concentrate on designing coils which will by themselves produce all the magnetic field that is required. This principle seems to have been adopted now and it is unlikely that any bigger magnets will be built for general magnetic research. For special purposes in which a moderate field is required over a large volume iron-cored electromagnets still hold their own and the world's largest magnets nowadays are those built for some specific purpose, the largest of which are those employed in particle accelerators. We shall defer description of these giant magnets until a later stage.

The initial problem in making a solenoid that is capable of producing large magnetic fields is to remove the heat generated in the windings. This heat is caused by the electrical resistance of the wire, and it is worth inquiring whether the problem can be overcome at its source by using wire of very low resistance. In the first place one tries to do this from the start by using relatively few turns of thick wire, since the resistance of a wire is inversely proportional to the area of its cross section. In practice copper wire is used, since although silver is a slightly better electrical conductor its expense usually rules it out as a practical material. There also remains the possibility of cooling the windings to a low temperature, for we may recollect that the electrical resistance of a metal decreases as its temperature is lowered. The decrease may not be very great unless the metal

---

* It may be difficult to get these figures in their correct perspective and since they are going to get larger in the next few pages it is useful to have some idea of what they mean. Power supplied at the rate of 1,000 kilowatts would heat sufficient water for a hot bath in about ten seconds. At the standard rate of one penny per kilowatt-hour, 1,000 kilowatts, supplied for one hour, would cost about £4.

Magnetism

is exceptionally pure, and so this imposes a further restriction on the wire used for winding the coil. It seems that the experimental difficulties of constructing useful magnets operating continuously at low temperatures are such that tangible improvements are attained only when the resistance of the coil is reduced to about one fiftieth of its resistance at room temperature. This involves cooling high purity copper to temperatures of the order of 50° K by the use of large quantities of liquid hydrogen, which naturally boils away rapidly when the magnet is switched on. Recently it has been demonstrated that supplies of extremely pure copper and liquid hydrogen are available in the U.S.A. in sufficient quantities to make low-temperature magnets a practical feasibility, and such magnets have been constructed operating in boiling hydrogen, producing a steady field of 60,000 oersteds acting over one or two cubic centimetres with a power dissipation of only 15 kilowatts. It is possible that these magnets will be developed still further in the future, and if the power saved in the magnet by operation at low temperature is greater than that used to liquify the hydrogen (for the gaseous hydrogen can be collected and liquified again by refrigeration) they may become economical. At present one of the limiting factors is the immense size of the refrigerator that would be needed to maintain the necessary continuous supply of liquid hydrogen.

The most successful solenoids for producing large magnetic fields which have been built up to the present time are those in which the heat is removed by water cooling. The pioneer in these experiments was F. Bitter who, shortly before World War II, designed a magnet in which the coil was made of a flat copper spiral with holes bored through it at regular intervals so that cooling water could be forced through them as well as round the spiral (see Plate 23). Bitter's largest magnet ran at 1,700 kilowatts and produced a field of 100,000 oersteds over a volume of 25 c.c. The technical difficulties of constructing a magnet such as this are enormous. The cooling water has to be pumped through the coil as quickly as possible, and since it is in immediate contact with the metal forming the coil it must be distilled, for ordinary tap water is a reasonable conductor

Magnetism in Scientific Research

of electricity and if this were used electrolytic corrosion of the coil would take place. Special electrical generators are required to provide the power for these magnets – Bitter's (Plate 22(b)) can supply 10,000 amperes at 170 volts continuously, which is equivalent to a power of 1·7 megawatts.* Since the war numerous high-power solenoids of rather more conventional design have been constructed in laboratories in various parts of the world. Usually they are made in sections and each coil consists of a spiral of thick copper tape insulated by nylon spacers (see Plate 24). Water under pressure is forced through the spirals to carry away the heat generated by the electric current. One of the best known solenoids of this type is that constructed by Daniels at Oxford, and is shown in Plate 30. The design and construction of solenoids such as these demands considerable care and skill, for the electromagnetic force of attraction between neighbouring coils due to the current they carry is very large. Moreover the heat generated is so great that an enormous flow of water has to be maintained – the cooling water exerts a force on the casing of Daniels' solenoid equivalent to a weight of nine tons – and several instances have been reported of solenoids being used well within their electrical power rating not because of an inadequate supply of electrical power but because a sufficient water pressure could not be maintained. Some examples of continuous-working high-power solenoids are shown in Plates 22, 24, 30, and 31(a).

In the nineteen-sixties a number of laboratories were set up to provide continuous high field facilities in excess of 100,000 oersteds. At the Royal Radar Establishment, G.B., a pair of co-axial solenoids, each with its own separate power supply, can be used to generate fields up to 180,000 oersteds. However the most ambitious and, to date, successful venture is the Francis Bitter National Magnet Laboratory at M.I.T. It uses four generators each capable of delivering 2·5 megawatts. By using a triple co-axial coil system, each separately energized, fields of up to 250,000 oersteds can be produced for short periods. To achieve this power is dissipated at the rate of 16 megawatts.

* 1 megawatt = one million watts.

Magnetism

A few sentences above we were careful to use the word 'continuous' for it so happens that physicists can often make do with large magnetic fields that last for only short periods of time. This opens up great possibilities, for not only is the heating much reduced if the current is maintained for only a short period, but furthermore a much greater momentary current can be delivered by the supply than is possible continuously. The production of large but momentary magnetic fields – 'pulsed fields' as they are usually termed – was pioneered by the Russian physicist P. Kapitza at Cambridge in the early nineteen-twenties. Basically the idea is to convert a large store of energy into magnetic energy as quickly as possible. One rather obvious way of doing this is to connect a large battery of lead accumulators to a low-resistance coil for a short period of time. This was the first method that Kapitza tried, and by this means he was able to obtain fields of about 400,000 oersteds lasting for a few thousandths of a second. There are certain drawbacks to this method. In the first place the lifetime of the accumulators is much reduced by being subject to such treatment, and secondly the switches have to be very complicated in order to start and stop the enormous currents at the appropriate time. These difficulties were never really overcome, and Kapitza eventually decided to use an A.C. generator as his source of supply. We may recall that the e.m.f. from such a generator varies with time in such a way as to be instantaneously zero twice every complete cycle. Kapitza's use of a generator was based on the idea of switching the current on at a time when the e.m.f. was instantaneously zero and switching it off at the next. In the meantime the full e.m.f. of the generator was applied to the magnet for one half cycle of its output and currents reaching 15,000 amperes could be delivered when the e.m.f. was at its peak. In this way Kapitza regularly achieved fields of 300,000 oersteds – only for short periods of time of course because the whole sequence of operation lasted only one hundredth of a second. Several difficulties were encountered. Coils carrying such large currents are subject to an enormous stress proportional to the square of the magnetic field they produce, and several of Kapitza's coils burst as a result. In addition the dynamo suffers an intense shock when it is short cir-

Magnetism in Scientific Research

cuited across the coil. We can understand this by reference to Lenz's Law, but it will in any case be realized that in this experiment the magnetic energy in the coil (also proportional to the square of the field strength) is obtained via an electric current, at the expense of the mechanical energy of rotation of the dynamo. When it is connected to the coil it is suddenly stopped and in doing so imparts an immense force to the floor (some idea of the magnitude of the force may perhaps be gained by trying to stop a child's humming top or a bicycle wheel suddenly). The shock travels to all parts of the building and in Kapitza's case it was more than sufficient to upset the delicate measuring apparatus he was using for his magnetic measurement. Kapitza's solution was elegant and simple. He placed the generator a long way from the coil in which the field was to be produced, so that his instruments had recorded the magnetic field and its effect before the shock from the dynamo, which travels at the speed of sound in the building, had reached them.

As the whole world knows, Kapitza was prevented from returning to Cambridge after his annual holiday in Russia by the Russian authorities in 1935, and although his magnet and generator were eventually sent to him in Russia where they eventually resumed working, Kapitza's departure so seriously interrupted further research into the production of enormous magnetic fields that it was not resumed until after the war. By this time advances in the design of electrical condensers and in the materials used for their construction opened up a new means of storing energy for subsequent conversion into magnetic fields.

If a condenser consisting of two parallel plates separated by an insulating material is connected to a high-voltage supply, the condenser becomes charged until the potential difference between its plates is equal to that of the supply voltage. The electrical energy contained by a charged condenser is proportional to its capacity and to the square of the voltage between its plates. We can regard a charged condenser as a battery which runs down very quickly when delivering an electric current. Consequently if we connect a coil of low resistance to a charged condenser the effect is very similar to that produced by connecting it to a large battery, but with the great advantage

Magnetism

that the condenser 'runs down' in a fraction of a second and consequently no switch is necessary to break the current. This method was actually tried by Kapitza, but it failed because the condensers of his time could not store enough electrical energy. Modern electrical condensers impose no such limitations, and the post-war attempts at producing large pulsed fields have almost invariably used them as the source of electrical energy.

Considerable progress has already been made and Kapitza's 300,000 oersteds has been surpassed on many occasions and in a number of different laboratories. The problems are chiefly the engineering ones of electrical circuitry and mechanical failure, for every time the field is trebled the pressure which the coils have to withstand due to their own magnetic field is increased ten-fold, and at 500,000 oersteds this pressure is 40 tons per square inch. Plate 27 shows the way in which a coil without adequate mechanical support has become distorted by the stress imposed upon it by the magnetic field. The type of equipment that has successfully and repeatedly been used to produce pulsed magnetic fields up to 750,000 oersteds is illustrated in Plates 25 and 26. In view of the tendency of spiral coils to buckle and deform, attempts have been made to use coils consisting of a single turn. Single turn means a cylinder of an alloy of beryllium and copper whose mechanical properties are inferior only to steel, one inch in external diameter and with a hole of diameter one-eighth of an inch bored through the centre. When connected to a suitable bank of charged condensers momentary currents of up to two million amperes have been produced. Fields of 1.6 million oersteds have been obtained in this way. In order to achieve this the currents are so large as to cause momentary melting of the metal. The magnetic force is likewise so great that the inner diameter of the coil which produced 1·6 million oersteds increased from $\frac{1}{8}$ to $\frac{5}{8}$ inches and the single turn which began as a short cylinder with a small hole through its centre ended as a long sheath less than a quarter of an inch thick (see Plates 28 and 29). This may give some idea of the difficulties which beset the production of very large magnetic fields. Some of them may be partly overcome, but there is still a long way to go before a field equivalent to the internal field inside a ferromagnetic substance is achieved.

Magnetism in Scientific Research

One of the earliest uses of large magnetic fields was in the production of extremely low temperatures. Liquid helium boils at 4·2° K at atmospheric pressure. The boiling point of a substance decreases with decreasing pressure, and to this is due the fact that water boils at a lower temperature at the top of a mountain, where the atmospheric pressure is less than at sea level. Consequently if the pressure in a vessel containing liquid helium is lowered, the liquid helium will boil and, taking its latent heat from its own heat energy, the temperature will fall. The lowest temperature which can be reached in this way is about 1° K, the limit being set by the speed with which the helium gas can be pumped away, and for a while this appeared to be the lowest temperature practically realizable. However in 1926 Debye, in Germany, and Giauque, in California, working independently of each other, both suggested the same method for reaching temperatures much lower than this. It is based upon the fact that when a substance is magnetized by the application of a magnetic field it warms up slightly and cools again when the field is removed. We can see in a general way why this should be so, for when a field is applied to a paramagnetic substance the magnetic moments tend to become aligned, and application of the principle of Le Chatelier (page 39) leads us to the conclusion that the substance reacts in such a way as to oppose the alignment, which it does by increasing its temperature. By good fortune the effect is much the largest with paramagnetic substances at low temperatures, for in these the greatest degree of alignment can be produced by a magnetic field (for this reason ferromagnetic substances are unsuitable, since their magnetic moments are already aligned by the internal field).

The technique employed is illustrated in Figure 45. The paramagnetic salt is contained in a thin-walled vessel containing helium gas and which is placed in a Dewar vessel (a large 'Thermos' flask) containing liquid helium boiling under reduced pressure at 1° K. The whole assembly is placed inside a further Dewar vessel containing liquid nitrogen to prevent heat from the room being communicated too rapidly to the liquid helium. At the beginning of the experiment the paramagnetic salt is at 1° K. When the magnetic field is switched

Magnetism

on the salt heats up, and this heat is transferred by the surrounding gas to the liquid helium outside, some of which boils away. Then the salt is thermally isolated by pumping away the helium gas in the container. The magnetic field is then switched off and the salt cools. Because of its thermal isolation no heat can reach it from outside and its low temperature is maintained.

The temperature which can be reached by this means depends

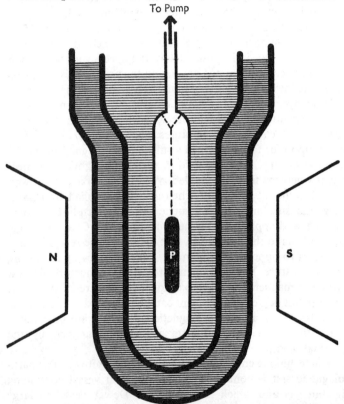

Figure 45. Magnetic cooling. The paramagnetic salt P is suspended inside a vessel which may be evacuated or filled with helium gas. The vessel is surrounded by liquid helium contained in a Dewar vessel which is itself surrounded by a Dewar containing liquid nitrogen.

The whole is placed between the poles of an electromagnet.

Magnetism in Scientific Research

not only on how well the paramagnetic substance is isolated from the outside but on the magnetic field (for this determines the extent of the alignment of the magnetic moment) and the choice of substance as well. Fields of about 30,000 oersteds are usually used. The ideal substance would be one which obeyed Curie's Law down to absolute zero. Since this would imply complete absence of interactions between the individual magnetic moments we cannot expect to find such a substance in practice. Those which obey Curie's Law most closely are the magnetically dilute salts mentioned on page 97 in connexion with paramagnetic saturation, and salts such as potassium chrome alum and gadolinium sulphate have been used successfully. It is by no means simple to measure the temperature attained by this process, since ordinary thermometers cease to be usable long before they are reached. One method is to estimate the temperature from the susceptibility of the salt, assuming that Curie's Law is obeyed. However it is impossible to verify the validity of Curie's Law and measure the temperature from a single measurement, and considerable ingenuity has been lavished on experiments to find exactly what temperatures have been attained. Usually these are in the region of $0.01°$ K but by using special modifications of the technique somewhat lower temperatures have been reached.

One example of the kind of information which can be obtained only by working at low temperatures is provided by work carried out at Oxford. The nucleus of an atom has a magnetic moment of its own. So far we have neglected to mention this, and we are justified in so doing by virtue of the fact that nuclear magnetic moments are about 2,000 times weaker than atomic moments and in consequence they make no significant contribution to the magnetic properties of matter that we have so far encountered. Nevertheless at very low temperatures it is possible to observe a small nuclear paramagnetism due to alignment of nuclear magnetic moments. Nuclear alignment should give us information concerning the nucleus and its structure just as the alignment of spin and orbital moments of the electrons provides us with information about the structure of atoms. There are several ways of in-

Magnetism

vestigating the properties of atomic nuclei, but the Oxford group chose the most straightforward method of using the artificial radioactivity of an isotope* of cobalt. When this substance decays its nuclei emit γ-rays, and if the nuclei are aligned by application of a strong magnetic field at very low temperatures it is observed that the γ-rays are emitted preferentially in certain directions. This indicates that the nucleus, in spite of its minuteness, has a definite structure, and by measuring the way in which the emitted γ-rays are distributed important information concerning its structure can be obtained.

The fact that atomic nuclei possess magnetic moments leads to many interesting and important consequences, one of which must be mentioned here. Some time after the first successful adiabatic demagnetization experiments it was suggested that, by demagnetizing the nuclear paramagnetism at about $0.01°$ K, even lower temperatures might be reached. It was estimated that a temperature of the order of one millionth of a degree absolute might be reached under favourable conditions. The experimental difficulties are immense and the initial experiments were unsuccessful. In 1955, however, a group at Oxford succeeded for the first time in demagnetizing first the electronic paramagnetism and then the nuclear paramagnetism. (Their apparatus is shown in Plate 30.) By this means they reached a temperature of about 15×10^{-6} ° K – the lowest temperature so far achieved by man.

Since the remainder of the applications of magnetism to physical research are going to be chiefly in the realm of atomic and nuclear physics it will first be necessary to explain what influence a magnetic field has on atomic particles. Actually it would have no effect at all were it not for the fact that most atomic particles are electrically charged. A few are neutral, such as the neutron and certain mesons (see page 226), and these are unaffected by a magnetic field.† A beam of charged

* See page 223.

† Some of these particles may possess a magnetic moment, in which case they may be aligned by a magnetic field. This effect is quite different from the bending of a stream of charged particles, which is the chief consideration of this section.

Magnetism in Scientific Research

particles consists of a quick succession of particles in rapid motion like bullets from a machine gun, and as such behaves like an electric current. It therefore experiences a force in a magnetic field which is the same as the 'motor effect' discovered by Faraday and whose direction is given by Fleming's Left-Hand Rule. But whereas with the motor effect the wire moves, the whole path of the particles is bent when they are travelling in a magnetic field. The force for a wire of length l, carrying a current i and placed at right angles to a magnetic field H, is Hil. The current is equal to the charge q passing in time t, so the force can be written Hql/t. In a single charge of magnitude e, we can write this expression as Hev, where v is its velocity, the direction of the force being always at right angles to both H and v. The path followed by a charged particle in a magnetic field is often very complicated, particularly if the field happens to vary from place to place. So we shall confine our attention to the case in which the magnetic field is uniform and the force on the particle is constant. A particle which moves under a constant force at right-angles to its direction of motion moves in an arc of a circle (see Figure 46). A stone attached to a piece of string and whirled above our head moves in a circular path because it is acted upon by a constant force, the tension in the string, which is always at right angles to its direction of motion.

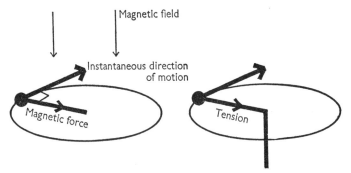

Figure 46. A particle which is acted upon by a force at right angles to its path moves in a circular arc. (a) Charged particle in a magnetic field. (b) Stone attached to a string.

Magnetism

Similarly, a charged particle moving at right angles to a uniform magnetic field always moves in a circular path, the radius of which may be determined from the requirement that the magnetic force on the particle Hev towards the centre must be equal to the centrifugal force mv^2/R, outwards. Thus

$$Hev = mv^2/R$$

or $$R = \frac{mv}{He}$$

The curvature imposed by the field on the path of the particle, $1/R$, is equal to $\frac{eH}{mv}$ so that the bending is proportional to the magnetic field strength and the charge on the particle, as we should expect; it is less for heavier particles than for light ones and is smaller the greater their velocity. Thus electrons are easily deflected, protons less so, while rapidly moving nuclei of the heavy elements are very difficult to deflect at all. The extent of the bending also depends on the size of the region over which the magnetic field extends, for if this is greater than R the particle will revolve in a complete circle, while if it is less the particle will be bent through an angle less than 360° so that it

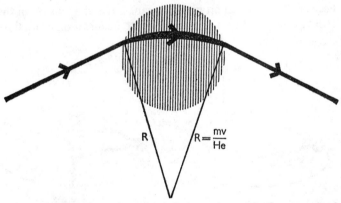

Figure 47. Deviation of a beam of charged particles by a magnetic field. Outside the field (shaded and acting at right angles to the plane of the diagram) the particles travel in a straight line. Within it they travel in a circular arc of radius mv/He.

Magnetism in Scientific Research

emerges from the magnetic field, whereafter it travels in a straight line since it is no longer acted on by a force. In this case we speak of deflection or deviation of the path of the beam, and by measuring the magnitude of the deviation it is possible to calculate the radius of curvature of its path in the magnetic field, as shown in Figure 47. The strength of the magnetic field can usually be calculated, and so, since R is known, the quantity mv/e can be found. Neither m nor v nor e can be determined separately by a single measurement of the magnetic deflection, and supplementary experiments have to be carried out if any one of these quantities needs to be measured. As we shall see this rarely happens and magnetic deflection alone can provide a vast amount of information of great value to the nuclear physicist.

The first use of this principle was made by J. J. Thomson in experiments which led to the discovery of electrons. As mentioned earlier (page 68) these were first found in electrical discharges in gases at low pressure, and it was by measuring the ratio of the charge on these particles to their mass that their nature was established. As we have seen magnetic deflection

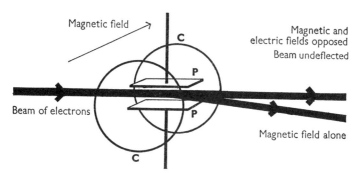

Figure 48. The measurement of the ratio of the charge to mass of an electron. A beam of electrons is bent upwards by the electric field between the condenser plates PP and downwards by the magnetic field acting between the pair of coils CC. By adjusting these fields the electric and magnetic deflections can be made to annul each other. From this the velocity of the electrons can be calculated. Deflection in a magnetic field alone then gives the ratio e/m.

Magnetism

alone gives only mv/e, and a separate experiment is needed to determine the velocity of the particles. This was done by passing the beam of particles not only through a magnetic field, but through an electric field as well, as shown in Figure 48. An electric field exerts a force of magnitude Ee on any charged particle whether moving or not, and by arranging for the electric field to be at right-angles to the magnetic field and both at right-angles to the beam the particles could be deflected in the same direction by both fields. Alternatively by suitably arranging the strength and direction of these fields the magnetic and electric deflections could be made to act in opposite directions and cancel each other out, so that the beam passed through them undeflected. In order that this may occur the electric and magnetic forces must be equal, i.e.:

$$Ee = Hev$$

so that $v = E/H$. J. J. Thomson found that the velocity of these particles was very high – about 3×10^9 cm./second (about 20,000 miles per second) or roughly one tenth of the velocity of light. Once v is known e/m could be determined from the magnetic deflection alone, and it was found that e/m always had the same value (about 1.8×10^7 electromagnetic units of charge per gram) whatever the gas originally present in the discharge tube. Soon afterwards it was established that the particles which are emitted when light or X-rays are allowed to fall on a metal surface (photo-electric effect, nowadays used in photographic exposure meters and television cameras), also possessed the same ratio of charge to mass. In a similar manner it was established that the particles given off by a heated wire (thermionic effect, which today forms the basis of wireless valves) and the β-rays emitted by certain radioactive substances could be identified with electrons, which thus appeared as a universal constituent of matter. Today we make extensive use of electrons and of the fact that they are deflected by electric and magnetic fields, for this forms the basis of the cathode ray tube used in oscilloscopes and in television receivers.

The electrons which J. J. Thomson discovered in the electrical discharge tube originate from the atoms within it, and every

Magnetism in Scientific Research

atom, in giving up an electron, becomes positively charged. Thomson realized that by measuring the ratio e/m possessed by these positively charged ions he could compare the masses of atoms directly and, as it were, determine the atomic weight of a substance one atom at a time. It is not necessary to know e since this can only be one, two, or perhaps three electron charges, and although some ambiguity may arise because one does not know which, it is always possible to perform experiments to overcome this. It is by no means such an easy task to measure e/m for charged atoms as it was for electrons, because their masses are several thousand times as great and the magnetic deflection correspondingly smaller. Consequently whereas a pair of coils sufficed for the magnetic deflection of electrons a large electromagnet was necessary for these positive ions. Eventually J. J. Thomson succeeded in deflecting them and one of his first experiments led to the discovery of isotopes. It was found that certain atoms, although possessing the same atomic number and hence, by virtue of the possession of the same number of electrons, identical chemical properties, could nevertheless exist in different forms, each one having a different mass. Such different forms of the same chemical element are termed isotopes. Neon for example turns out to be a mixture of two isotopes of mass 20 and 22, of which the former is the most abundant in nature. The discovery of isotopes explained why chemical atomic weights of many elements rarely come out to be a whole number. Such substances are a mixture of isotopes each one of which has an isotopic weight which is a whole number, and the chemical atomic weight is the average of the isotopic weights of its isotope, weighted in proportion to their relative abundance.

Thomson's work was continued and extended by Aston in England and by several other physicists in the United States. Their instruments, known as mass-spectrometers, showed the existence of a very large number of isotopes (something like 400 are now known), and by careful measurement established that the isotopic masses were themselves not exactly whole numbers. Accurate determination of the actual masses of the atomic nuclei (for the mass of the electron is small and can be

Magnetism

allowed for) forms the basis of any calculation of the energy available from nuclear reactors for power stations or other devices making use of nuclear power.

It will be realized that since the isotopes of an element have identical chemical properties, it is impossible to separate them by chemical means. Instead, use has to be made of the very slight differences in physical properties which do exist, and by means of which partial separation of the isotopes of an element may be obtained. In order to obtain a perfect separation it is possible to use the mass spectrometer, for this essentially sorts out charged particles according to their different masses by means of the magnetic deflection. The quantities of matter which may be separated in this way are however exceedingly small, but after running an instrument for a sufficiently long time it was found possible to obtain enough of the two isotopes of lithium to take measurements on them separately.

The difference in the masses of the two lithium isotopes is quite great, for they are in the ratio of six to seven and hence the magnetic deflection of the two isotopes are in the same ratio, roughly 1·16 to one. During the war, in the early days of the atomic energy project, a problem of an altogether different magnitude presented itself. The fission of uranium had been discovered in 1939, but it was later pointed out by Bohr that only the isotope of mass 235 would undergo fission with slow neutrons, and that in order to make a successful bomb this isotope, which constitutes only 0·7 per cent of all uranium atoms, would have to be separated from the isotope of mass 238. The first method successfully employed made use of the principle of magnetic deflection. The technical difficulties must have been immense, for the amounts required were kilograms rather than micrograms, which was the best that had been obtained from the separation of the lithium isotopes. On account of their large masses, the magnetic deflections are very small indeed and the difference between them are likewise small. Consequently in order to achieve appreciable separation not only were very large fields necessary, but they also had to extend over a large area so as to make the magnetic deflection as large as possible. It is known that many large magnets were

Magnetism in Scientific Research

used for this separation and that sufficient U_{235} was produced by this means to manufacture atomic bombs.

An example of the use of magnetic deflection of atomic particles which gives sufficient information by itself is provided by the β-ray spectrometer. β-rays are electrons emitted from the nuclei of radioactive atoms. They are emitted with a wide range of velocities, and by measuring these we can find out a great deal about conditions inside the nucleus. Now since β-rays are electrons their e/m ratio is constant and known. Consequently determination of their radius of curvature in a magnetic field alone tells us how their velocities vary amongst themselves. The rays emitted from a radioactive atom are bent into circles by a magnetic field and are received on a photographic plate. When this is developed a number of black lines are seen to correspond to points where the β-rays have fallen after having been bent into a semi-circle. There are as many lines as there are different velocities, and by measuring their position, and knowing the magnetic field applied, the velocities of all the particles can be measured. A similar arrangement can be used to measure the energy of γ-rays by replacing the radioactive substance by a metal plate and allowing them to fall upon it. Electrons are liberated from the plate by the photo-electric effect, the energies of which are proportional to the energies of the incident γ-rays, and these may be determined by measuring their velocities from the magnetic deflection. The information obtained from these experiments was of fundamental importance. It led to concrete ideas concerning the nucleus and its structure and showed conclusively that the nucleus, like the atom, could exist in certain discrete energy states only. The results of the β-ray experiment required for their explanation the existence of a hitherto unsuspected particle much smaller in mass than the electron, and uncharged. This particle was termed a neutrino and, although the necessity of its existence was accepted as long ago as 1931 from the β-ray measurements, it was not until 1956 that its presence was detected by direct means.

Another example of the sufficiency of magnetic deflection is provided by research into cosmic rays. These are particles having enormous energies which reach the earth, apparently

Magnetism

from outer space. By an ingenious device known as the Wilson Cloud Chamber the track of these particles can be seen and photographed and by observation of their radius of curvature in a magnetic field one can measure their momentum (mass × velocity), since their charge is known to be numerically equal to that of the electron. One of the earliest observations was the existence of tracks clearly identifiable as being due to electrons but equally clearly being bent in the wrong direction by the magnetic field. This suggested that they were positively charged, and subsequent research demonstrated that these new particles, though resembling electrons in all other respects, carry a charge which is equal and opposite to that of the electron. They are called positrons and they are of great importance. Their existence, which had been predicted by Dirac on purely theoretical grounds before they were discovered experimentally by Blackett in England and almost simultaneously by Anderson in California, lends support to the idea of the symmetry of the particles which together make up the universe and strengthens the belief that every atomic and nuclear particle has its 'negative' counterpart. The discovery of the negative proton did not come until 1955 however, nearly a quarter of a century after the discovery of the positron. Both discoveries however were made possible only by the use of large magnets.

More recently a new set of particles has been discovered in the cosmic radiation. They are termed mesons and derive their name from the fact that their masses are intermediate between that of an electron and that of deuterium (heavy hydrogen of mass 2). They arrive at the earth's surface with enormously high energies, so high in fact that their maximum energies are not known, owing to the impossibility of imposing any measurable curvature on their tracks by the magnetic fields which we can make. How they acquire such energies remains one of the great mysteries of modern physics, and one of the interesting experiments planned for the future is to attempt to measure the energies of the fastest-moving particles, for this will resolve one further mystery, namely whether they come from our own galaxy or originate from extra-galactic space.

We now turn to an application of a slightly different kind –

Magnetism in Scientific Research

the use of magnetism as a means of endowing atomic particles with very high energies. Rutherford first disintegrated the atomic nucleus and brought about the first artificial transmutation of the element in 1919. Since then the desire to probe even deeper into the structure of the nucleus has increased steadily, and this has been greatly intensified by the discovery of a large number of mesons in cosmic radiation. Now we can only look into an atomic nucleus as it were through the agency of another atomic particle, and in order to find out what it is like inside a nucleus the particle must get inside or at least as close to it as possible. Then by observing how the nucleus behaves when another particle enters it or by observing how an atomic particle is affected by going very close to it we can draw conclusions about the nucleus and the forces which hold it together. Unfortunately the nucleus is positively charged, and, with the exception of electrons which are of little use for this type of work, all the atomic particles which we would normally use to probe the nucleus are positively charged too. Consequently there is a very strong repulsive force between the charges carried by the probing particle and the nucleus under investigation which prevents them from ever coming into close proximity unless the kinetic energy of the incoming particle is large enough to overcome it. The whole problem of atom splitting thus resolves itself into that of finding a beam of swiftly moving particles. Rutherford in his original experiment used the α-particles* emitted from certain natural radioactive substances and their energy set a limit to the number of nuclear transmutations he could induce. Obviously some means of accelerating the particles to a high energy is required. This can be done by placing them in an electric field where they experience a force and thereby acquire energy. However the electric fields which would be needed to produce particles with the energy desired correspond to potential differences of several million volts, and although this method has been successfully employed a limit is imposed on the energies attainable by the potential difference which ordinary insulators will withstand before breaking down.

A radically different method of accelerating particles was

* Helium nuclei – charge 2, mass 4.

Magnetism

suggested by E. O. Lawrence in California in 1932. It makes use of the fact that a particle is accelerated every time it passes through an electric field (provided this acts in the right direction), so that if on emerging from an electric field it can be made to change direction and pass through the electric field again in the same direction, then on emerging a second time its energy will be doubled. This can be done because we can alter the path of a charged particle by applying a suitable magnetic field. The principle behind Lawrence's device, which he termed a cyclotron, is the circular path which charged particles acquire in a magnetic field. The time for a complete revolution of the particle in the field can be calculated quite easily, for the distance travelled is $2\pi R$, its velocity is v, and so the time taken is $2\pi R/v$, which using the value for R given on page 220 comes out to be $2\pi m/He$. We see that this does not depend upon the speed of the particles nor upon the radius of its circular path.

In the cyclotron the particles to be accelerated are made to

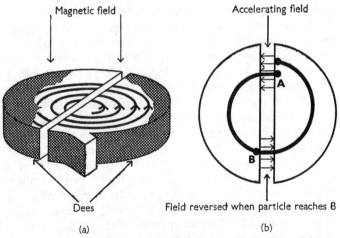

(a) (b)

Figure 49. The principle of the cyclotron. The charged particles are contained within the hollow dees. Within each dee the electrical potential is constant and there is no electric field. Since the dees themselves are at different potentials an electric field exists only in the gap between them.

Magnetism in Scientific Research

travel within a split metal box, rather like a circular biscuit tin which has been sawn into two semi-circular halves, as shown in Figure 49a. On account of their shape the two halves are usually known as dees. The dees are placed between the poles of a powerful electromagnet so that the magnetic lines of force are perpendicular to them, thereby ensuring that the particles within travel in circles. To understand how the cyclotron works consider Figure 49b and imagine a charged particle at A. If an electric potential difference is maintained between the two dees, an electric field exists in the gap separating them. Within the dees themselves the potential is constant and there is no electric field. In the electric field the particle will experience a force which therefore moves across the gap gaining speed as it does so until it reaches the edge of the other dee. Once inside this its speed remains constant, but being in a magnetic field it pursues a semi-circular path until it reaches B. If by the time it reaches B the potential difference between the two dees has been reversed, then the electric field it encounters at B will once again accelerate it across the gap. After this it will move in a circle of slightly larger radius, so that if the changing potential difference between the dees can be made to keep in step with the particle it will travel in a spiral path, gaining speed every time it crosses the dees until it reaches their outer edge. Here it is arranged to encounter a strong electric field which gives it a sudden deflection out of the dees, so that it may be used to bombard other atoms. By this means the particle acquires the same energy as if it had been accelerated by a potential difference equal to that across the dees multiplied by the number of transits made, and since a particle may make many thousands of revolutions before reaching the outside of the dees it is clear that this can easily be equivalent to many million volts.

The potential across the dees can conveniently be made to reverse its direction by applying to them an alternating voltage of the same frequency as the frequency of revolution of a particle within them. Frequency is the number of revolutions per second which is the inverse of the time for one revolution; thus $f = \dfrac{He}{2\pi m}$. For a proton, e/m is about 10^4, and assuming H to be

Magnetism

15,000 oersteds, which is a typical figure for a large machine, we find that f is roughly 2×10^7 cycles/sec. or 20 mc/s, which is a typical frequency for short-wave radio transmission. The required frequency is therefore not impossibly high, although it is obviously necessary to apply as large a potential as possible across the dees, and the difficulties of providing a really high voltage at such a frequency are quite considerable. These difficulties were overcome, and Lawrence successfully designed and built a cyclotron which accelerated particles to an energy equivalent to several million volts.

It would seem at first sight that the energy imparted to a particle depends only on the number of revolutions it can be made to undergo, and since this in turn depends only on the radius of the dees, the only limit to the energy of the particles could be imposed by the sheer physical size of the machine. Lawrence's first cyclotron was quite small by present-day standards, with pole pieces a mere foot in diameter. Subsequently larger machines were deemed necessary and shortly after the war Lawrence constructed a new one, with pole pieces sixteen feet across, designed to give an acceleration equivalent to a hundred million volts. The magnet required for this weighed just over 4,000 tons. Bearing this in mind it will be realized that the cost of these machines both in construction and operation is immense, and the real limit to the particle energy obtainable from them is imposed more by financial consideration than any other. To make matters worse serious physical difficulties appear in the way of the acceleration of particles to energies of more than about a hundred million volts. One of the predictions of the theory of relativity is that the mass of a particle increases with its velocity. The increase is very small except for velocities approaching that of light. When this happens to a particle inside a cyclotron its time of revolution is no longer independent of its speed, and consequently it gets out of step with the constant frequency of the accelerating potential. The kinetic energy of a particle is $\frac{1}{2}mv^2$ and so for a given energy the particle having the smallest mass is that with the greatest speed. For this reason electrons cannot be accelerated to high energies by a cyclotron, for with their

Magnetism in Scientific Research

small mass they rapidly acquire high velocities, their mass increases, and they cease to gain energy owing to the lack of synchronism between their motion and the accelerating potential.

Various methods have been suggested to overcome this difficulty. One way is to increase the magnetic field towards the outer edge of the dees in such a way that the ratio H/m remains constant. Another is to allow the particles to be emitted in bursts, and to vary the frequency of the accelerating potential so that it always keeps in step with the rotation of the particles. This is the usual procedure, and it will be appreciated that it adds considerably to the complexity of the instrument. Such machines are known by various names – synchrotron, proton synchrotron, or frequency modulated cyclotron. The first of these to be constructed was at the University of Birmingham shortly after World War II, but owing to shortage of material and labour it was not actually completed until one had been constructed in the U.S.A. Since then others have been built, one of the largest being at the University of California (Plate 21). The magnet for this machine weighs 10,000 tons, and stands about 14 feet high and is over 130 feet in diameter. A similar machine at Brookhaven cost just over seven million dollars to build. In order to obtain the fullest use from these machines it is necessary to keep them in operation for twenty-four hours a day, and this in itself costs the Brookhaven laboratories about a million dollars a year. Even larger machines are now being built, each requiring an even larger magnet than its predecessor. With machines such as these, particles can be accelerated to energies comparable with those possessed by cosmic ray particles, and their behaviour can be studied under closely controlled conditions. In this way the negative proton was discovered (at California), and further systematic studies will no doubt provide a key to the mysteries of the nucleus and nuclear particles which at present appear stranger as every new fact is discovered.

The cyclotron and its progeny are used only for accelerating the heavy particles encountered in modern physics and for reasons already discussed cannot be used to accelerate elec-

Magnetism

trons. A means of doing this was developed in America during the early years of the war. This device is called a betatron (Plate 31(b)), and in addition to using a magnetic field to impose a circular path on the electrons, it accelerates them in a simple but highly ingenious manner. Its operation depends on the principle stated by Maxwell and mentioned in Chapter 3 (page 43) that a changing magnetic field always produces an electric field whose lines of force are circles round the lines of force of the changing field. This electric field can be used to accelerate electrons, since they are already obliged to travel in circles by the magnetic field. A great deal of complicated mathematical analysis was necessary before this principle became an accomplished fact, and this was not achieved until 1941. Basically the betatron consists of a magnet whose field is rapidly increased as the electrons to be accelerated are injected into the gap between its poles. As the magnetic field increases the electric field thereby created accelerates the electrons and they spiral outwards just like heavier particles in a synchroton, gaining energy as they do so, until they emerge from the edge, where they can be used for whatever purpose they are intended. Since the mass of the electron is so small the curvature imposed on them by a magnetic field is much greater than for heavy particles, and betatrons are much smaller devices than cyclotrons. The energy to which the particles can be accelerated depends once again upon the size of the magnet, but electron energies of a hundred million volts can be produced by a betatron only a few feet across. They are frequently used nowadays for the production of very short wave X-rays used for deep X-ray therapy in hospitals.

Another device which makes use of the circular motion of electrons in a magnetic field is the magnetron (Plate 32), which is used for generating electromagnetic waves of very short wavelength such as those used in radar. Conventional oscillators can be used to generate these waves but become grossly inefficient as the wavelength decreases, and the development of the much more efficient magnetron arose from the necessity of equipping aircraft with radar for navigational purposes during the war. It consists essentially of a heated filament from which electrons

Magnetism in Scientific Research

are emitted, placed in an evacuated tube which is situated in a magnetic field. In the presence of the field the electron paths become circular and, by means of small cavities bored in the outer part of the tube, the circular motion of the electrons is able to set up electrical oscillations within those cavities which, if connected to a suitable aerial, make it emit electrical waves of very short wavelength. In much of the radar equipment this wavelength was 10 cm., although shorter wavelengths were used later. The magnetic field for a magnetron is provided by a permanent magnet, the field employed ranging from one to five thousand oersteds, depending upon the wavelength required. The great advantage of the magnetron is that it can produce extremely powerful bursts of these waves, thereby endowing radar with a much greater range than it would have if a less powerful oscillator were employed continuously.

Our first example of the use of magnetism in research was the production of low temperatures and our last is appropriately enough going to be the production of high temperatures. If an alternating magnetic field is applied to a metal the changing magnetic flux sets up eddy-currents within it which cause the metal to become hot. The temperature attained depends not only upon the strength of the magnetic flux but on its rate of change, i.e. on its frequency. Heating of metals in this way – induction heating as it is called – is accomplished by placing them inside a coil carrying a strong high-frequency alternating current. After quite a short time sufficient heat is generated to melt the metal and this provides a very convenient way of making alloys in small quantities.

The temperatures needed to melt metals and even the most refractory substances rarely exceeds a few thousand degrees. Now the temperature of a gas is a measure of the speed of random movements of its molecules, and at very high temperatures these so-called thermal speeds become comparable with those of atomic particles accelerated by a cyclotron. Consequently there is a possibility of bringing about nuclear reactions merely by maintaining a gas at a very high temperature, so that collision between rapidly moving particles happens very frequently. In some of these nuclear reactions energy is

Magnetism

evolved which can be used to maintain the temperature of the gas. Thus there is the possibility of a self-sustaining process in which the temperature of the gas is maintained by nuclear reactions within it. This mechanism is believed to be responsible for the heat of the sun, and since the raw material is chiefly hydrogen it is clear that if a process of this kind could be reproduced on the earth on a much smaller scale a virtually unlimited source of energy would be available.

The chief difficulty is that the temperatures required to initiate such reactions are about ten million degrees, much higher than any which are normally encountered in the laboratory, and, as pointed out earlier, the technical difficulties of reaching such a temperature are considerably greater than those of attaining a temperature of 10^{-3} ° K. One of the most serious troubles is that of containing the hot gas, for no solid can exist at such temperatures, and so at all costs it must be kept from the walls of the containing vessel. This can be achieved by magnetic means, for at temperatures such as these all gases are ionized and, being charged, are deflected by a magnetic field. Thus the hot gas may be 'contained' by a magnetic field of suitable shape. The high temperatures themselves may be produced by making use of the transformer principle.

CHAPTER 13

The Earth's Magnetism

WE shall devote the final pages of this book to a brief account of the earth's magnetism and its effects. The existence of the earth's magnetism is very important, for it is this which makes possible the use of the compass for navigation. Moreover it has more recently been found to be connected with radio transmission – also of great importance to ships at sea. Consequently a great deal of factual information has been accumulated concerning the earth's magnetism, its magnetic field, and the manner of its variation from place to place on the earth's surface. The collection of much of this information was begun a very long time ago, so that by 1830 Sir George Airy, the Astronomer Royal of that day, was able to construct a chart showing how the earth's magnetic field varied over the extreme northern hemisphere. Thus the magnetic field of the earth was the first to be thoroughly investigated experimentally, and the problem of its origin may be regarded as one of the earliest magnetic problems systematically studied.

In an earlier chapter we have described how Peregrinus mapped out the magnetic lines on a sphere of lodestone from observations that a pivoted compass needle placed on its surface set along certain lines, which if joined up formed meridians on its surface. He also observed that the lodestone sphere, if floated in water, always set itself north and south. William Gilbert repeated and extended Peregrinus's observations and reasoned that if a compass needle set itself in certain directions on the surface of the lodestone, and the lodestone set itself in a certain direction, namely, north and south, then the earth itself must be a magnetized sphere. This excellent discovery later proved to be a very useful one, for the earth's magnetic field varies from place to place in a manner very similar to what would be expected if the earth were uniformly magnetized. Moreover the magnetic field outside a uniformly magnetized

Magnetism

sphere is exactly equivalent to that which would exist if all its magnetism were concentrated in a magnetic dipole placed at its centre and pointing in the same direction as the sphere's magnetization. The magnetic field of a dipole may be calculated, and we may recall that it depends only on the strength of the dipole moment and the distance from it. In the early part of the nineteenth century Gauss showed that the existence of a magnetic dipole of suitable magnitude and direction, near but not quite at the earth's centre, would give rise to a magnetic field which agreed closely with that actually observed (see Figure 50).

In order to specify the characteristics of the earth's magnetic field at its surface we must remember that a magnetic field is a

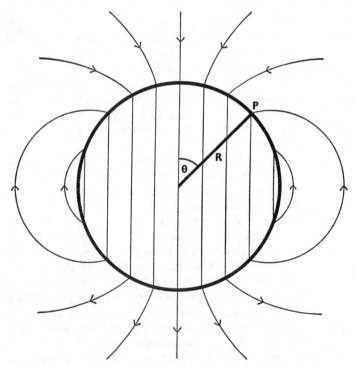

Figure 50. The magnetic field of a uniformly magnetized sphere.

The Earth's Magnetism

vector quantity and requires three quantities for its description. It does not matter which three quantities are given provided that it is possible to determine from them both the absolute magnitude of the field and its direction in space. Usually this is done by specifying the vector components of the field in the horizontal and vertical directions and also the direction of the horizontal component in a horizontal plane. These quantities,

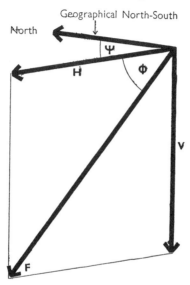

Figure 51. The magnetic elements, corresponding roughly to the P on Figure 50. F = Total field; H = Horizontal component; V = Vertical component. These are related by the law for vector addition, namely by the relation $F^2 = H^2 + V^2$. ϕ = Angle of dip; ψ = Declination.

which are illustrated in Figure 51 and which should be noted in connexion with Figure 50, are known as the magnetic elements. Another quantity which is often used in describing the earth's field is the angle of dip. This is the angle between the direction of the total magnetic field and the horizontal. It can be measured by taking a compass needle, pivoted about a horizontal axis, and placing it along the direction of the horizontal

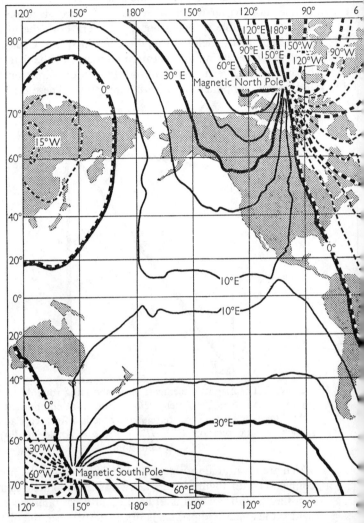

Figure 52. An isogonic chart. Along each line

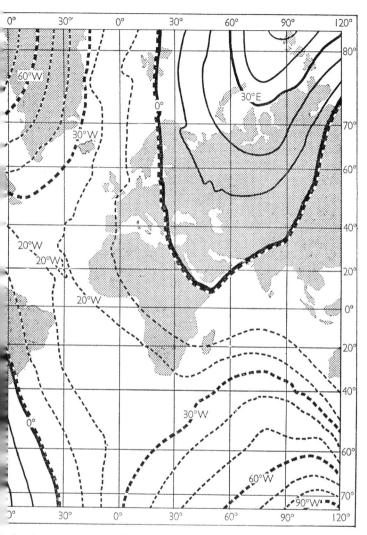

clination is constant and marked in degrees east or west.

Magnetism

component. It then sets itself along the line of force due to the earth's field, and the angle it makes with the horizontal can be measured. No new information is given by the dip since the angle ϕ is given by $\tan \phi = V/H$. At the equator the angle of dip is zero and the field is entirely horizontal, whereas at the poles the field is vertical and the horizontal component is zero. This provides a means of locating the magnetic poles, for the usual form of compass is pivoted so as to rotate in a horizontal plane only. It is thus acted on only by the horizontal component of the earth's field, and if this is zero it will show no preference for pointing in any particular direction. In this way Amundsen was able to locate the position of the magnetic north pole in 1904. The fact that the magnetic north pole does not coincide with the geographic north pole is of great importance to mariners and explains why it is necessary to specify the direction in which the horizontal component of the earth's field acts. The angle between this direction and geographic north is known as the declination. The declination varies considerably over the earth's surface and it is obviously very important to know this accurately, for only then can the reading of the compass, which points magnetic north, be converted to show which direction is geographical north. For this purpose accurate measurements of the declination are made at many points over the earth's surface and from these charts are prepared on which places of equal declination are joined by lines. These lines of equal declination are termed isogonic lines and the map on which they are drawn, an isogonic chart. From this and an ordinary compass reading, a navigator, provided he knows his position roughly, can determine the geographic north and set his ship accordingly.

A typical isogonic chart is shown in Figure 52. As one might expect, the declination is greatest near the poles themselves and decreases as the distance from the poles is increased. Records show that the declination has changed appreciably over the last century or so and is presumably changing still. For example, whereas in 1815 the declination at London was about 24° w, it is now only about 11° w. We shall refer to this at a later stage, but one consequence is immediately obvious – the

The Earth's Magnetism

changes are so great that isogonic charts must be revised and brought up to date at intervals of a few years if they are to be of any use for reliable navigation.

The next task is to see how well the hypothesis of the earth being a uniformly magnetized sphere accounts for its field and its variation over the earth's surface. The declination must be due to the fact that the equivalent dipole at the earth's centre does not coincide in direction with the geographic axis of the earth's rotation. If we denote the strength of the dipole by M and the radius of the earth by R then the horizontal and vertical

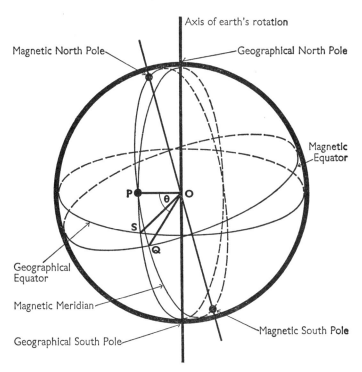

Figure 53. Diagram showing magnetic poles, meridian, and equator and their relation to the corresponding geographical quantities. The magnetic latitude at P (cf. Figure 50) is the angle POQ. Its geographical latitude is the angle POS.

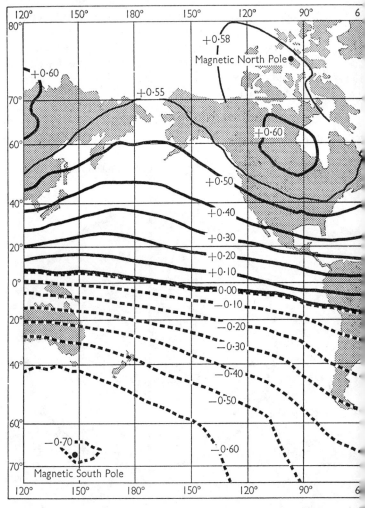

Figure 54. Magnetic chart of equal vertical intensity. Along each
like form of these lines at the equator due to the non-coincide

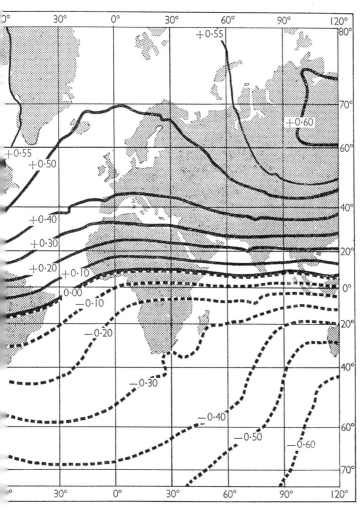

vertical component of the field, V, is constant. Note the wave-
the magnetic and geographical equators.

Magnetism

components of the field due to this dipole at the earth's surface are

$$H = \frac{M \cos \theta}{R^3}$$

$$V = \frac{2M \sin \theta}{R^3}$$

where θ is an angle which we shall call the magnetic latitude. If the axis of the magnetic dipole coincided exactly with that of the earth's rotation the declination would be zero and the magnetic and geographical latitudes would be identical. The difference is slight (Figure 53), but it means that in order to test the magnetized-sphere theory of the earth's field we must specify locations by their magnetic latitudes rather than by their geographic ones.

It turns out that in order to account for the values of the earth's magnetic field which are actually observed M has to be $8 \cdot 06 \times 10^{25}$ and, since R is 6370 kilometres or $6 \cdot 37 \times 10^8$ cm., M/R^3 comes out to be $0 \cdot 312$ oersteds. This means that at the magnetic equator, where $\theta = 0$, $V = 0$ and $H = 0 \cdot 312$ oersteds, and at the north and south magnetic poles, where $\theta = 90°$, $H = 0$ and $V = 0 \cdot 624$ oersteds. Thus from the equator to the north pole the horizontal component decreases steadily from its maximum value to zero, and at the same time the vertical component increases from zero to its maximum value. Both fields are independent of magnetic longitudes, but because the magnetic and geographical equators do not coincide they do depend slightly on geographical longitude. These features are brought out most clearly in Figure 54, which is a chart showing the lines of equal vertical intensity. An additional check is provided by calculations of the angle of dip. For this angle ϕ is given by $\tan \phi = V/H$ and using the expressions obtained above for V and H it is evident that

$$\tan \phi = 2 \tan \theta$$

which allows the angle of dip to be calculated from the magnetic latitude. At London this is about 54°, and from our formulae we can calculate H, V, and ϕ to be $0 \cdot 138$, $0 \cdot 505$ oersteds, and 70° respectively. These values are to be compared with the observed values (1950) of $0 \cdot 186$, $0 \cdot 433$, and $66\frac{1}{2}°$. Similar agree-

The Earth's Magnetism

ment, good but not quite exact, would be obtained for these magnetic quantities at all other places.

Thus we see that the uniformly-magnetized sphere hypothesis accounts very satisfactorily for the general features of the terrestrial magnetic field and its variation with position. There are certain slight discrepancies between theory and experiment. For example the lines of equal vertical intensity are not smooth but show kinks and irregularities at certain places, and the same is true of the isogonic lines which ought, according to theory, to be much more regular. Most of these differences can be satisfactorily accounted for by a number of subsidiary factors which we have so far ignored. In the first place the earth is not spherical but is appreciably flattened at the poles. Secondly, we can hardly expect the earth to be uniformly magnetized throughout. Thirdly, we have always to reckon with the possibility of local irregularities due to the presence of magnetic ores close to the earth's surface. These are by no means inconsiderable, and indeed the local disturbances of the earth's field which they cause can be used to locate oil and minerals by carrying out detailed aerial magnetic surveys.

Much more difficult to explain is the fact that the earth's magnetism also varies with time, an observation which has already been mentioned in connexion with isogonic charts. These variations are known as secular changes, that is, slow changes continued in the same sense though not necessarily at a constant rate. The existence of the secular variation was first discovered by Henry Gellibrand at London in 1634, the declination itself having been known at least since the fifteenth century. During the seventeenth century the secular variation of the angle of dip was noted, and Figure 55 shows the variations of both at London since 1580. The curve is an oval and suggests a periodic variation of about 500 years. Variations observed at other places tend to show different periods, and it is doubted whether the overall secular changes for the whole earth are periodic at all. Secular changes in the horizontal and vertical components have been observed, but it is only within the last hundred years that determinations have been made which are sufficiently accurate to show them clearly. The indications are

Magnetism

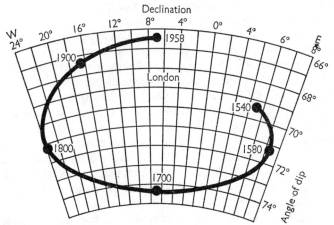

Figure 55. Secular variation of the declination and dip at London since 1580.

that the earth's field has decreased by about 5 per cent over the last hundred years. Methods of inferring the declination and strength of the earth's field in historic and prehistoric times are dealt with in the next chapter.

In addition to the secular variation careful measurement has revealed the existence of much more rapid changes. These changes are very much smaller than the secular changes, but are none the less of great importance. They appear to originate from effects outside the earth whereas the secular changes arise from within. The type of variation usually observed is a daily one. Naturally the variation is greater some days than others, but it always conforms to a daily pattern. There is in existence an international agreement to classify days into 'quiet' and 'disturbed' days according to the extent of the magnetic variation, and Figure 56 shows a typical example of the quiet day variation. The daily variation at all places depends upon the local time (that is, the position of the sun in the heavens), and this, together with the fact that the amplitude of the variation is invariably greater in summer than in winter, suggests very strongly that the sun is primarily responsible. It was suggested

The Earth's Magnetism

as long ago as 1882 by Balfour Stewart that these diurnal variations might be caused by horizontal electric currents circulating in the upper atmosphere at right-angles to the earth's vertical field and thus giving rise to a magnetic field at the earth's surface by the ordinary laws of electromagnetism. About forty years later Appleton discovered the existence of electrically conducting layers in the earth's upper atmosphere. There are several of these, but our chief concern at present is that the ionosphere, which they collectively constitute, contains a very large number of ionized atoms and molecules, which endow the ionosphere with its electrical conductivity. The degree of ionization and the thickness of the ionosphere are determined by a number of factors, one of which is the strength of the ultra-violet light rays from the sun. If these ionized layers are set in motion, perhaps by a solar tidal effect or by thermal connexion or gravitational effects, the motion will

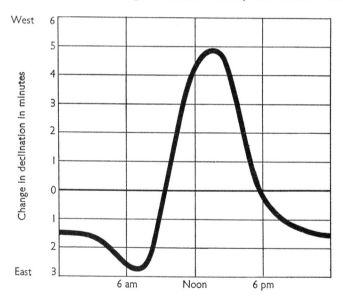

Figure 56. Typical daily variation of the declination on a 'quiet day'. The variations of the other magnetic elements from their average values are of the same general type.

Magnetism

eventually resolve itself into a circular one round the earth, and a magnetic field will be set up proportional to the amount of movement and to the number of ions present.

This explains the general character of the variation observed, for we should on this basis expect it to depend upon the amount of sunlight received and agree with the observed dependence on local time (due to the earth's rotation about its axis) and with the seasonal variation (due to the rotation of the earth round the sun). It also explains another fact not previously mentioned, that in addition to the diurnal variation there exists also a semi-diurnal variation which itself varies regularly in magnitude with a period of twenty-eight days. These additional variations, which comprise only a small part of the total diurnal variation, are almost certainly due to currents set up in the ionosphere by the moon as a result of tidal action similar to that which causes the sea tides twice every lunar day.

The circumstantial evidence for attributing the diurnal variations to causes outside the earth and in the ionosphere in particular is thus very great, and it is confirmed by observations of a quite different kind. When radio waves enter the ionosphere from the earth they are deviated by it much in the same way that light is bent by a glass prism, or when entering water at an oblique angle. This is known as refraction, and the high refractive index of the ionosphere for radio waves is due to the presence of the electrically charged ions which are present. Were it not for this ionosphere, long-range radio reception would be impossible, for radio-waves, like light waves from which they differ only in wavelength, travel in straight lines and thus are rapidly lost from the earth as a result of its curvature unless transmitted from a very high aerial. The ionosphere is able to bend some of the radio waves to such an extent that they are reflected down again to the earth's surface. The refractive index of the ionosphere, which determines the degree of the bending, depends on the wavelength of the radio waves and for wavelengths corresponding to television frequencies is insufficient to reflect them. Consequently they travel outwards and are lost to the earth, with the familiar result that transmission of television takes place only over short distances. With wavelengths of the

The Earth's Magnetism

order of 30 metres (frequency 10 mc/s) which is about five times those used for television purposes, the deviation is sufficiently great to bend the radio waves down again. These wavelengths are those commonly used for short-wave radio transmission, and as is well known short-wave radio propagation extends over very large distances indeed. Short-wave reception conditions are notoriously variable and show a diurnal variation which has all the characteristics associated with the diurnal variation in the earth's magnetic field. It would thus seem that ionospheric currents can satisfactorily account for them both, and this view is supported by detailed investigations of the two occurrences.

A further interesting feature is brought to light by a careful study of the daily variation on 'quiet days', for if this is averaged for a whole year (to take care of the fact that the variation is greater in summer than in winter) it is found that this average varies periodically with a period of about eleven years. Now an eleven-year period is frequently found in connexion with many terrestrial events. Sometimes the periodicity in events which is claimed is based more on speculation than on evidence, but there are good grounds for the belief that periodic climatological changes occur with an eleven-year period. Evidence for this comes not only from recent meteorological records but also from the fact that the growth rings on certain long-lived trees vary in the same manner. It is however the sun which shows the eleven-year period most strikingly and it is likely that all the terrestrial changes which vary with the same period do so because of events taking place in the sun.

The events referred to are connected with the sunspot activity of the sun. Examination of the sun's surface shows a large number of dark spots. Usually these can only be seen with the aid of a telescope, but sometimes they are so large that they can be seen with the naked eye by looking at the sun through dark glass or at sunset. These sunspots seem to move over the sun's surface and appear and disappear in an irregular manner. If however the number of sunspots in evidence is averaged over the whole year, this average is found to vary with an eleven-year period. Moreover there is a very striking correlation

Magnetism

between the daily 'range' (i.e. the size of the daily variation) in the horizontal component of the earth's magnetic field averaged over a year and the average number of sunspots, as is shown in Figure 57. It is likely that the sun spot activity has a great effect on the amount of ultra-violet light reaching the earth. This influences the depth of the ionosphere and the degree of ionization within it, thereby affecting the diurnal variation in the magnetic field. Moreover since plant growth is stimulated by ultra-violet light it is evident that the periodicity in the growth rings on trees can be attributed to the same cause.

In addition to the magnetically 'quiet days', some days exhibit much more irregular variations and are known as 'disturbed'. If the degree of disturbance is very great the term

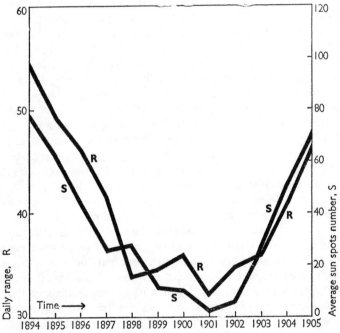

Figure 57. The relation between the daily range, i.e. the extent of the daily variations on 'quiet days' averaged over a whole year, and the average number of sun spots.

The Earth's Magnetism

magnetic storm is used to describe it. The irregular variation on disturbed days also shows a very strong correlation with sunspot activity, as Figure 58 shows. The study of magnetic storms has revealed further interesting facts which go some way towards an explanation of their origin.

One of their most striking features is that they usually commence suddenly, at almost the same instant all over the earth; in really intense storms the differences between the time of their beginning recorded at different stations seems to be no more than about half a minute. Furthermore it is found that magnetic storms tend to recur at intervals of about twenty-seven

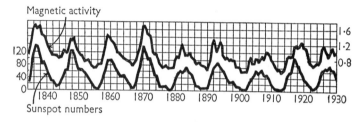

Figure 58. The daily variation on 'disturbed days' averaged over a year and the average sunspot activity.

days. This is the time taken for the sun to revolve once about its own axis, indicating that one particular sun-spot or group of sun-spots is responsible for each magnetic storm, and is effective only when facing the earth. As the sun rotates, the solar disturbance misses the earth until it has made one complete revolution, when it is effectively in the same position again and produces another magnetic storm. The third feature of interest is that there appears to be a definite time lag of between one and four days between an increase in sun-spot activity and the occurrence of magnetic storms on the earth. Light from the sun reaches us in about eight minutes and this suggests that the sun-spots, or something associated with them, emit particles. It is now believed that the particles originate not from the sun-spot itself but from the so-called solar flares. These intense areas of light which occur near sun-spot areas usually start

251

Magnetism

suddenly, often lasting only for a few minutes and rarely longer than an hour. The flares themselves seem to be composed mainly of atomic hydrogen. In order to reach the earth in the time suggested by the observations the particles emitted must travel at speeds between 270 and 1,100 miles a second. We do not know what these particles are, although their origin strongly suggests that they must be protons, but we may be quite certain that they are electrically charged. When they enter the ionosphere they increase the degree of ionization already present and thereby produce a magnetic storm. These storms are invariably coupled with changes in short-wave radio transmission conditions. During a magnetic storm long distance transmission becomes very uncertain and severe magnetic storms may be accompanied by a complete radio fade-out. Such fade-outs temporarily disrupt radio communication and have occasionally been known to affect cable telegraphy. In this respect they are a nuisance, but from them we can learn about the nature of magnetic storms and indirectly about the sun as well.

The problem of explaining the earth's field has now been reduced to that of accounting for the main field and its secular variation. What might at first sight seem to be difficult to explain – the short-term variations – turns out to be easily understandable in principle, even though the exact details may be lacking. The really difficult problem is the original one, that of the main magnetic field. This field could be adequately explained if the earth were a uniformly magnetized sphere. Unfortunately it is not, for its intensity of magnetization would have to be 0·075 gauss in order to account for the magnitude of the observed field, and it is quite certain, magnetic rocks notwithstanding, that the earth's crust is not magnetized on the average to anything like this extent. In the first place most rocks are very feebly magnetic, with susceptibilities comparable with those of paramagnetic substances, and secondly, such magnetism that they do possess decreases with increasing temperature and thus steadily diminishes as we go downwards into the earth's crust.

A second possibility is that there does actually exist a huge

The Earth's Magnetism

magnetic dipole at the earth's centre. We recall that this dipole moment has to be as large as $8 \cdot 06 \times 10^{25}$ in order to account for the field. Now the magnetic moment of the dipole is intensity of magnetization times volume, and as we saw in an earlier chapter the maximum intensity of magnetization never seems to exceed about 2,000 gauss. So the dipole would have to occupy a volume of about 4×10^{22} cubic centimetres, which corresponds to a sphere whose radius is about one thirtieth of the whole radius of the earth. This seems not unreasonable at first, but there are other factors which rule it out as a practical possibility. The most important of these is the fact that the temperature increases rapidly towards the centre of the earth, which in consequence has a liquid core. All ferromagnetic substances lose their ferromagnetic characteristics at the Curie temperature, and for iron this occurs at 770° C, a temperature attained at a depth of only sixty miles. The inside of the earth is however subject to enormous pressure owing to the contraction of the cooled outer layer, and it has been suggested that since the Curie point depends on pressure, conditions at the centre of the earth may allow iron to be ferromagnetic in spite of its high temperature. Unfortunately laboratory experiments show that the Curie temperature of iron actually decreases with pressure, and this hypothesis, which always had an air of desperation about it, is at present considered to be untenable.

If the earth's magnetism cannot be attributed to the presence of permanent magnetism within it the next logical step is to inquire whether electric currents circulating within the earth's core could exist in such a way as provide the magnetic field. One of the chief difficulties here arises from the fact that the currents have to be maintained continuously; there must be a source of electromotive force present, as otherwise any currents would decay owing to the electrical resistance of the core. A theory which provides a possible mechanism for this has been developed by W. M. Elsasser in America and by Sir Edward Bullard in England. They suggest that the earth behaves somewhat like a self-exciting dynamo, which produces a magnetic field by means of electric currents induced by that field in the core.

Magnetism

Some energy has to be provided in one form or other to maintain the dynamo, which would otherwise appear to have all the characteristics of a perpetual-motion machine, and it is suggested that this comes from thermal convection within the core (which is believed to consist mainly of molten iron) which arises from the heat developed by radio-active changes, and which is superimposed on the normal circular motion of the core owing to the rotation of the earth. A simple self-exciting dynamo which serves to illustrate the basic idea behind the theory is shown in Figure 59. A metal disc rotates about an axis and its rim and axle are connected as shown to a stationary coil whose axis coincides with the axle. If a small magnetic field exists in the direction of the axle, an e.m.f. will be induced between rim and axle and a current will flow through the coil, which if suitably connected will produce a field which reinforces the original field. The field thus increases until the heat generated by the current in the coil is equal to that supplied to drive

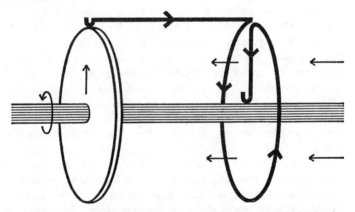

Figure 59. Simple self-excited dynamo. If a small magnetic field exists in the direction shown, then clockwise rotation of the disc (looking in the direction of the field) produces an e.m.f. from axle to rim. If a coil is connected by sliding contacts to rim and axle as shown, the current in the coil produces a magnetic field which is in the same direction as the original field. The principle is the same as the series-wound dynamo in electrical engineering.

The Earth's Magnetism

the wheel. Conditions inside the earth are vastly more complicated than this because, whereas the dynamo of Figure 59 is essentially a two-dimensional one involving rotation about a single axis and with the current flow determined by the coil, no such limitation is imposed by the spherical core of the earth. Consequently the shapes of the possible magnetic fields and the types of current flow which provide them become very complicated indeed. Nevertheless the dynamo theory has far fewer objections than most others which have been put forward (and these are very numerous indeed), and whatever modifications may be necessary before it gains final acceptance it seems likely that the proposed mechanism or something closely akin to it will finally prove to be the correct one.

The existence of the earth's magnetism has not unnaturally led to some speculation as to whether other heavenly bodies may not have a magnetic field. Intuitively this seems likely, since we have little grounds for believing our own planet to be unique. In practice it is not at all easy to detect magnetism elsewhere than on the earth. Direct measurement of any field which may be presumed to exist is out of the question on account of its certain minuteness at the earth's surface. The moon is by far the nearest to the earth of all heavenly bodies, being only some 240,000 miles distance. This is roughly 4×10^{10} cm., and using this value for R in the equations on page 244 we find that the earth's dipole moment would give rise to a field of only 2×10^{-6} oersteds on the surface of the moon. This is much too small to be detected, and similar calculations show that the field of the planets and the sun and all other heavenly bodies would be, even in the most favourable circumstance, negligible in comparison with the irregular changes in the earth's own field.

It is worthwhile digressing here somewhat to explain how the magnetic fields of the sun and of some stars have not only been detected but measured. The measurement is based on observation of the light emitted by the stars themselves, and in this sense carries the measuring device into the star itself. When atoms emit light they do so at certain definite wavelengths just in the same way that a radio or television transmitter

Magnetism

sends out waves of one wavelength only. In the presence of a magnetic field each single wavelength that the atom would normally emit is split up into a group consisting of a small number of different wavelengths centred around the original. The separation between the wavelengths split up in this way is proportional to the magnetic field strength in which the atom is situated. This effect was looked for by Faraday, and although he was unsuccessful in his search we can only marvel that he should have suspected this close connexion between magnetism and light. It was eventually found by the Dutch physicist Pieter Zeeman towards the end of the last century and it is nowadays usually known as the Zeeman effect in his honour. By allowing the light to fall on a prism or on a similar device which separates it out into its various colours or wavelengths, the separation of the wavelengths can be measured, from which the field strength acting upon the emitting atoms may be deduced. In practice this is much more difficult to carry out than simple statement suggests. The wavelength separation is never very great, nor is the amount of light which even the biggest telescopes can gather from distant stars. Consequently, although it is necessary to use an arrangement which separates the colours as much as possible, there is a danger that the colours so dispersed will be too faint to be observed.

It was therefore recognized as a great achievement when the American astronomer Hale detected the sun's magnetism and found that there exists a magnetic field at its surface of about fifty oersteds. Unfortunately Hale's observations have not been confirmed by others, and it is possible that the field is considerably smaller than this. All astronomers agree however in finding a large Zeeman effect in the light emitted from sunspots, corresponding to magnetic fields as great as 3,000 oersteds. These observations give support to the view that a sunspot is a region of violent whirling motion of the ionized gas that is known to exist on the sun's surface. Since the war another American astronomer, Babcock, using a greatly refined measuring technique, has succeeded in measuring the Zeeman effect in the light emitted from a number of stars. Sur-

The Earth's Magnetism

face fields as great as 8,000 oersteds have been deduced from some of these observations.

It is more than likely that all stars have magnetic fields of their own, and it was suggested by Blackett, reviving an earlier hypothesis by Schuster, that the magnetic fields of the earth and the stars may be a fundamental property of rotating bodies, which is to say that any rotating body possesses a magnetic moment by virtue of its rotation alone. This suggestion has the great virtue of simplicity; but this simplicity is in fact being bought at a great price for it is tantamount to the assumption of a new law of nature, and physicists, who strive to explain all physical phenomena in terms of the smallest possible number of hypotheses, are naturally reluctant to admit the existence of an additional law of nature until they find themselves faced with no alternative. Present experimental evidence is against Blackett's hypothesis. Babcock found from his Zeeman effect measurement that some stars possess a varying magnetic field; one even reverses at regular intervals without the accompaniment of a corresponding change in the sense of its rotation. Some useful information can be gained by measuring the magnetic moment of the moon, for the moon is solid throughout and the electrical mechanisms which have been proposed for the earth's magnetism cannot account for any magnetism the moon may possess. This was one of the tasks which was successfully carried out by the Russian rocket Lunik II. It confirmed conclusions inferred from indirect evidence by two astronomers, one in Russia and one in France. The moon, which has no atmosphere and can support no life, has no magnetic field either.

CHAPTER 14

Rock Magnetism

THE earliest magnet known to man was a piece of rock from the earth, and the magnetism of rocks has been appreciated and put to practical use for a very long time. Indeed, within a few decades of Gilbert's discovery of the earth's magnetism, certain strongly magnetic ores, chiefly deposits of magnetite, had been located because of the local distortion of the earth's field which they produced. Instruments such as the Swedish Mining Compass, invented by Daniel Tilas in 1672, were devised specially for this purpose, and the use of magnetic observations is probably the oldest method of geophysical prospecting used in the search of economic mineral deposits. As is only to be expected, such instruments were crude and insensitive and they were effective only in searching for strongly magnetic deposits. Within this century magnetometers, as they are called, have been greatly improved and their sensitivity increased so that aerial surveys can detect local changes in magnetic field as little as 10^{-5} oersteds, thereby locating ores which are only weakly magnetic, while laboratory instruments have been constructed which will respond to a magnetic field change of as little as 10^{-9} oersteds. With the aid of these instruments it has been established that almost all rocks are magnetized. Part of a rock's magnetization is of course due to the fact that it is situated in the earth's field, but when allowance is made for this it is always found to be permanently magnetized to a certain extent. Often the magnetization is extremely feeble, but it does exist and moreover can be measured with considerable accuracy. These facts have been known for some time, but it is only very recently that the magnetism of rocks has been sufficiently studied systematically for its significance to be appreciated. The whole subject is thus in its infancy, and it is not intended in this chapter to do more than touch lightly upon a few of the discoveries which have been made and on their possible interpretation.

Rock Magnetism

The magnetism of rocks is due to the presence of minute quantities of minerals which are almost invariably ferrimagnetic. Truly ferromagnetic fractions are probably very rare indeed. These minerals are present in the form of fine grains, in consequence of which the rocks themselves, although weakly magnetic, are frequently magnetically hard with coercive forces which may reach several hundreds of oersteds. When we come to inquire how such a rock acquires its present magnetization we can immediately recognize the significance of any magnetism it may possess. For a rock, in common with all other substances, can become magnetized only by being exposed to a magnetic field and the magnetic field operative here is that of the earth itself. Consequently by studying the magnetism of its rocks, we can hope to discover information which will tell us about the earth's field in prehistoric times when the rocks were being laid down. Any changes in the earth's field which have taken place in geological time should be permanently enshrined in the strata which the geologist uncovers.

The mechanism whereby a rock becomes permanently magnetized depends to a great extent on the type of rock itself and this in turn depends on how the rock has been formed. Igneous rocks are those which have been ejected from the earth's interior in the molten state and have solidified on being deposited on the earth's surface. The most obvious examples are the lava flows produced by erupting volcanoes. Such rocks acquire their magnetization by being cooled from above their Curie temperature to normal temperatures in the presence of the earth's magnetic field. When this field is removed the rocks are found to possess a permanent remanent magnetization known as the thermo-remanent magnetization. This is very much greater than that which they would have acquired by being subject to the same magnetizing field at ordinary temperatures, and it is retained much more tenaciously.

The magnetization of an igneous rock is therefore not only quite appreciable, but is, in addition, very stable. This is extremely fortunate because it means that rocks may be collected during field expeditions and brought back to the laboratory, where they may be investigated using sensitive laboratory

Magnetism

techniques, with the assurance that their properties will not be significantly different from those which they possessed *in situ*. The magnetization of such rocks can at least in principle tell us the direction of the field which was acting on them when they cooled through their Curie temperature. Bricks and pottery become magnetized in the same manner after firing, and consequently can be made to yield information about the earth's field in historic times.

Sedimentary rocks are those which have been formed from fine particles produced from igneous rocks by weather erosion. Having become suspended in the water of lakes, rivers, and seas they eventually sink to the bottom and become compacted by pressure into solid rocks. In the formation of sedimentary rocks it seems likely that the magnetic constituents, being suspended and therefore free to rotate, become aligned in the earth's field and consequently become deposited with their magnetic moments directed along the earth's field in existence at the time. As might be expected the magnetization of sedimentary rocks is very much weaker than that of igneous rocks, since there have been no high temperatures to assist the acquisition of magnetization. In addition it is found that such rocks are usually magnetically less stable than igneous rocks.

Rock magnetism can be studied from a number of different aspects. The presence of magnetic constituents in the rocks themselves presents a problem to the geologist. The nature of these constituents can only be decided from chemical and physical examination. The mechanisms whereby a rock becomes magnetized is itself a problem of great physical interest. Most of these are too specialized to find a place here and we shall refer only to the geomagnetic aspect, namely the possibility of interpreting the magnetization possessed by a rock in terms of the direction of the earth's field which was operative at the time at which the magnetization was induced.

The earth's field in historic times has been investigated by observation of the magnetism of historically documented lava flows, from human artefacts such as bricks, and from sediments. The first systematic investigation was carried out by Chevallier

Rock Magnetism

in France, who, in 1925, made a thorough study of the magnetization of the lava flows on Mount Etna, the eruptions of which are dated back to 1284. He was able to show that the direction in which the rocks were magnetized agreed with the known secular variation of the declination of the earth's field from about 1600 onwards, thus showing that this use and interpretation of the rocks' magnetism is basically sound. This impression was confirmed by the demonstration that the Yoridai-Sawa lava flow, deposited after an eruption on 12 July 1940, in Japan, became magnetized in the direction of the local magnetic field. More recently Thellier, also in France, investigated the magnetization of bricks taken from the walls of kilns excavated in Carthage which had apparently been in use up to the city's destruction. It was concluded from these investigations that the secular variation of the earth's field was at that time very similar to that of the present day. Recent measurements from Iceland indicate that between two and three thousand years ago the north pole was located east of the geographical pole, roughly at the tip of northern Norway. While this must await confirmation from other sources it is of great interest because it implies that the Greeks must have seen the 'northern lights' considerably more frequently than today, on account of the greater proximity of the magnetic north pole. This would explain Aristotle's great familiarity with the aurora as deduced from his exhaustive description of it in his *Meteorologia*. The earth's field, apart from undergoing the secular variation in declination, seems from these observations to have decreased in magnitude, and in Roman times was about one and a half times as great as it is now. This is in accordance with the comparatively recent observation that the earth's field is decreasing steadily (page 246).

Dating of rocks and bricks from historical records becomes virtually impossible before about 2000 B.C., and other methods have to be used to provide the time scale. This is provided by the so-called varved clay beds found in Sweden and New England. A varve is a bed of clay consisting of a sediment brought to the bed of a lake by streams originating near the tops of ice and snow-covered mountains. Each spring and summer as the ice thaws the lake receives a supply of sand and clay from the streams. The coarse material settles almost at once, but the finer particles remain in suspension much longer. In the autumn and spring both lake and glacial streams become frozen over and the fine particles settle gradually in the still water beneath, forming a thin layer of dark clay, easily distinguished from the thick layer of sand beneath. Each year the sand and clay is sorted out in this way, and thus a varve, comprising the two distinct layers, is laid down each year. The strata consists of numerous thin parallel layers, each set one year older than that immediately above it. The dating of the last Ice Age period by counting varves is one of the classics of geological research, but our concern here is that measurement of the direction of magnetization within each varve gives information about the direction of the earth's field at the time of their formation. Measurements have already been carried out on varves which have been dated back to approximately 10,000 B.C., that is to the last phase of the last Ice Age. It has been established that in the last 12,000 years or so the earth's field has been that corresponding to a magnetic dipole at the centre, with a secular variation in declination of between 30° east and west of the geographical north pole.

Twelve thousand years represents only an infinitesimal fraction of geological time, and the success of the magnetic method as applied to recent rocks led to attempts to study the magnetization of older rocks with a similar purpose in view. Interpretation here is considerably more difficult because of the obscuration of geological formations by more recent covering and by its total obliteration in certain cases by erosion. Great assistance can be rendered by the geologist, who from his own investigations is able to indicate those rocks whose ages can be ascertained with some certainty. Two very important results have already been obtained from the older rocks. In the first place the direction of magnetization does not correspond even approximately to the present direction of the earth's field, even when allowance is made for the secular variation; secondly rocks are frequently found to be magnetized in a reverse direction to that of the present field.

Rock Magnetism

The occurrence of reverse magnetization in rocks undoubtedly exists, although its interpretation is somewhat in doubt. When samples of the same type of rock are taken from a given locality they are invariably found to be magnetized along a certain direction and in the opposite direction, but not in directions intermediate between the two. Moreover normal and reversed magnetizations seem to occur with equal probabilities. The immediate interpretation to be placed on these results is that the earth's field has reversed its direction frequently in geological time. The absence of rocks magnetized in intermediate directions implies that these reversals must have occurred quite suddenly. There is no possibility of a slow shift, but rather an outright 'flopping over' of the poles which by geological standards must have been instantaneous. Unfortunately there are other possible explanations. Rocks are complex substances chemically, and the processes by which the igneous ones acquire their magnetization is complex physically, so much so that Néel has been able to think of four different physico-chemico mechanisms enabling an igneous rock to be magnetized in a direction opposite to that of the applied field. This interpretation received considerable support from the discovery, in Japan, of a rock which, when cooled in a magnetic field, acquires a magnetization which reverses on further cooling. On the other hand the widespread discovery of reversals of magnetization in sedimentary rocks has revived belief in the hypothesis of reversals of the earth's field. Opinion is still very much divided and is likely to remain so. There is nothing in the currently accepted dynamo theories of the earth's field to indicate that reversals are impossible or even unlikely to occur at fairly frequent intervals (for it has been estimated that several hundred reversals may have taken place in geological time). Theorists, however, have for the most part preferred to wait until the mechanisms proposed by Néel have been conclusively proved to be inoperative before attempting to explain both the field and its reversals.

The fact that the older rocks are frequently found to be magnetized in a direction significantly different from that of the present field of the earth is of the utmost importance because it

Magnetism

implies that at the time of their formation the pole positions were appreciably different from those they occupy at present. Geologists have often attempted to discover the physical geography of the earth in the past, that is, the past distribution of continents and oceans and the variations in climate. Unfortunately the geological record is often incomplete and sometimes obliterated altogether, so that many differing interpretations have been based on the geological information obtained. In particular there are two issues upon which opinion has been most seriously divided. These are the questions of polar wandering and continental drift. Many geologists believe that although many of the present land-masses may at one time have been covered by seas, and that continents may have been changed in form by the formation of new mountain ranges, the deep ocean basins and the land areas have occupied essentially the same position that they do today. Moreover, they hold that geological evidence provides no basis for the belief that the axis of rotation of the earth has ever been different in the past from that at present. Others maintain that the geological information reveals climatic changes in the past which can be explained only by assuming that the geographic axis of the earth has moved with respect to its crust, so that what are now polar regions might once have been equatorial and vice versa. Moreover, they argue that the configuration of the continents and the existence of certain similarities in their geological structure indicates that they were once joined together and have gradually drifted apart. This celebrated and controversial hypothesis of continental drift seems first to have been suggested more than a hundred years ago, but it is usually associated with the name of Wegener, who in a book published in 1915 marshalled an imposing collection of facts in its support.

In view of the fact that geological evidence alone seems to be unable to decide upon the correctness of these two hypotheses, independent evidence would seem to be helpful and this can to a certain extent be provided by magnetic observations once certain basic assumptions are admitted. The first of these is that a rock becomes magnetized in the direction of the existing

Rock Magnetism

magnetic field* and the second is that, ignoring the secular variation, the earth's field always coincides with the axis of the earth's rotation. The evidence for this is based on the magnetism of recent rocks, described earlier, which certainly obey these rules. Obviously no direct check on the validity of these assumptions is possible for the older rocks but indirect evidence in their favour is that, once they are made, the magnetic observations do show systematic variations with geological time.

It is only within the last five years or so that magnetic observations have been carried out in a really systematic manner, and although certain definite conclusions have already been drawn from these, there remains the possibility that they may have to be modified in the light of further discoveries. First is that the hypothesis of polar wandering seems to be supported. Figure 90 shows how the position of the north pole has varied, based on the directions of magnetization observed in European rocks. The dating of these rocks is fairly certain because only those whose age could be estimated reliably from geological considerations were used for magnetic measurements. The question which immediately arises is whether the pole positions given in Figure 60 are consistent with the geological record. To a very considerable extent they are. The climate of any region is in general determined by the intensity of the sunlight it receives and this is largely determined by its latitude. There is abundant geological evidence for the view that Britain in past geological time had a much warmer climate than now. One finds coal measures, coral reefs, and salt deposits characteristic of tropical climates. Particularly impressive are the large coal measures, which have always been regarded by geologists as having been laid down in regions of high temperatures and heavy rainfall. The coal beds are believed to have been formed in the Carboniferous period about 300 million years ago. Figure 60 shows that the north pole at this time was situated somewhere near the northern coast of China, and this implies that the greater part of Europe

* There is always the possibility of reversed magnetization, but since normal and reversed magnetizations occur equally there is rarely any ambiguity as a result.

Magnetism

lay quite close to the equator. Equally significant too is the fact that similar coalfields exist in the eastern part of the United States, which according to this position of the pole would also be experiencing a tropical climate at this period. To take another example: magnetic observations on certain Cretaceous rocks (150 million years) in Tasmania show that they are magnetized with a very steep dip, suggesting that Tasmania was at the time very near the south pole. The climatic evidence from this region suggests cool, even glacial conditions, during this period. Further examples could be

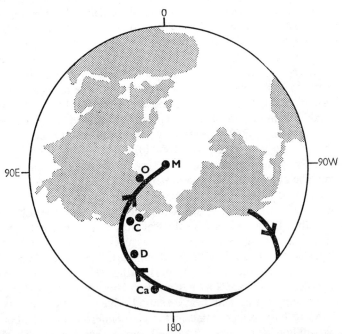

Figure 60. The position of the magnetic north pole in the geological past, deduced from observations on magnetic rocks collected in Europe. The thick line gives the approximate path of the polar movement. M, Miocene (35 million years old); O, Oligocene (50 million); C, Carboniferous (280 million); D, Devonian (320 million); Ca, Cambrian (500 million).

Rock Magnetism

given where the magnetic evidence agrees with the geological. It is only fair to add that certain examples can also be found where the evidence conflicts, but the hypothesis of polar wandering seems to be strengthened as every further investigation is reported.

Another finding which promises great significance is the following. Prior to Miocene times (about thirty-five million years ago) the pole positions deduced from all rocks of the same age are the same no matter from what part of the world the rocks are taken. With older rocks this is no longer the case, and in particular the position of the poles obtained from samples from continents in the northern and southern hemispheres differ appreciably. Further investigation has begun to reveal a further interesting feature, namely that whereas rocks of the same age and taken from the same continent are usually magnetized in the same direction, this direction differs from that of rocks of the same period from other continents. This strongly implies that the continents themselves must have moved with respect to each other since their rocks become magnetized. This of course is what the theory of continental drift predicts. Now the next question is: if the continents are imagined to be twisted back to their original directions so that their directions of magnetization all point towards the same pole position,* do they then fit together in the manner suggested by Wegener? Once again the magnetic observations suggest that they do.

The adherents of the continental drift hypothesis are themselves divided as to the origin of the continents. Wegener himself suggested that they were all originally joined together into a single 'Pangaea'. Others imagine two primeval continents, 'Laurasia' in the northern hemisphere and 'Gondwanaland' in the south, comprising the present land masses of South America, Africa, the Indian Peninsular, Arabia, Australia, and Antarctica.

Opinions differ as to the precise manner in which those land masses once constituted Gondwanaland, since various reconstructions are possible. One due to Du Toit is shown in Figure

* Note that on account of polar wandering this does not coincide with the present pole position.

Magnetism

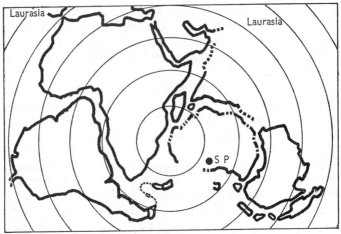

Figure 61. Du Toit's construction of Gondwanaland.

61. If this is correct then peninsular India should have moved northwards about 5,000 miles since partition occurred and rotated about 28° anticlockwise. Recently a number of different observations on igneous rocks in India have been made. They all indicate substantial northward movements during the last 100 million years since the rocks were formed. Rocks from Brazil and from Central Africa have also been investigated and are found to be magnetized in a direction consistent with the view that the continents of South America and Africa have also shifted and rotated during this period from positions which agree roughly with Du Toit's map.

In the northern hemisphere the magnetic results indicate that North America has also moved during this period from a position very much closer to Europe and Africa than it now occupies. If this were so then the Atlantic Ocean must have been considerably smaller in the distant past than it is at present.

To say any more at the present stage would be unfair to those who are working in this field, for this work is still very much in its infancy and some interpretations may have to be retracted and replaced by others. Moreover no useful purpose is gained in overstating the case for new ideas and interpretations. That

Rock Magnetism

Wegener's own hypothesis failed to achieve the universal acceptance that he had hoped was due in no small way to the exaggerated claims he made on its behalf. The interpretation of the magnetization of certain rocks leads us to suppose that continental drift has occurred. The evidence is at present by no means conclusive but it does, at present, seem to support the geological evidence for continental drift. Many workers believe however that further study of the magnetism of rocks from different parts of the globe will either confirm these preliminary findings and thereby establish the truth of the theory of continental drift or else it will be refuted entirely. If confirmed it should be possible to reconstruct the primeval continents of Laurasia and Gondwanaland, or of Pangaea, with considerable accuracy from magnetic observations alone. If the theory is refuted then other explanations will have to be sought for the geological similarities which exist between parts of the east coast of South America and the west coast of Africa. Explanations will have to be sought for the distribution of the same animal species in continents separated by vast tracts of sea. On the other hand if it is confirmed one is faced with the great problem of accounting for the forces great enough to cause widespread movement of large land masses. In either case the results and the conclusions to be drawn from them can hardly fail to be of interest not only to geologists and biologists but to all who are interested in science in its broadest sense.

Bibliography

Anderson, J. C.: *Magnetism and Magnetic Materials*. An introductory text book at the university undergraduate level.

Morrish, Allen H.: *Physical Principles of Magnetism*. A more advanced book than Anderson's.

Tebble, R. S.: *Magnetic Domains*. A brief account of magnetic domains at the undergraduate level.

Whittaker, E. T.: *A History of the Theories of Aether and Electricity*. The standard history of the subject, in two volumes, dealing with the development up to 1926.

Acknowledgments

IN writing this book I have drawn freely on existing books, articles, and technical reports and my first task is to express a debt of gratitude to all those authors (especially those who sent reprints) whose works I have used as sources. Likewise, I am obliged to a large group of persons, too numerous to name individually, who, in one way or another, have contributed both directly and indirectly to much that this book contains.

If specific mention of individuals who have, perhaps unwittingly, contributed to the text would be invidious, this is not so of those persons and organizations who have so readily supplied material for the illustrations. Plate 6 is reproduced by courtesy of A. D. Little Inc. Photographs for Plates 7, 8, and 9 were kindly supplied by the General Electric Company and those for Plates 10 and 11 by the Plessey Company Ltd. The Bitter patterns of Plate 12 originate from the laboratory of Professor L. F. Bates, whilst those of Plate 13 were kindly lent by Dr D. H. Martin. The examples of the use of permanent magnets, Plates 14(b) and 15 were offered by the Permanent Magnet Association and I am indebted to Mr J. E. Gould for obtaining them for me. The large magnets shown in Plates 16 and 20 are both in the Bellevue Laboratories of the Centre National de la Recherche Scientifique and I am grateful to the C.N.R.S. and its director, M. G. Depouy, for supplying me with the photographs. The smaller magnets shown in Plates 18 and 19 are those of the Newport Instrument Company and the Mullard Research Laboratories respectively. The photograph of the Bevation electromagnet appears through the courtesy of the University of California Lawrence Radiation Laboratory. The equipment shown in Plates 22, 25, 26, 27, and 31(a) is that in use at the Lincoln Laboratory of the Massachusetts Institute of Technology, the photographs being kindly provided by Dr H. H. Kolm. The excellent pictures shown in Plates 23, 24, and 30 are of magnets used at the Clarendon Laboratory, Oxford. They were taken by Mr C. Band and kindly supplied by Mr M. F. Wood. The illustrations of the single-turn magnets shown in Plates 28 and 29 are due to Dr R. W. Waniek. The photographs of the Betatron, Plate 31(b),

Acknowledgments

and magnetron, Plate 32, are by courtesy of the Metropolitan Vickers Electrical Co. Ltd and the M-O Valve Company Ltd respectively.

In addition thanks are due to the editors of *Endeavour* for permission to reproduce Plate 14(a) and Fig. 35; to the publishers of *Encyclopaedia Britannica* for Fig. 17; to Messrs Methuen and Co. Ltd for Figs. 52, 54, 55, 57, and 58; to the editors of *Philosophical Magazine* for Fig. 60; and to Messrs Oliver and Boyd for Fig. 61.

Lastly I would like to record my gratitude to my wife who by drawing my attention to many obscurities in the original version has contributed in no small way to whatever clarity of exposition the book possesses.

Index

acceleration of atomic particles, 227–32
AIRY, Sir George, 235
Alcomax, 179
alloys (*see also* steel)
  antiferromagnetic, 202
  ferromagnetic, 115–17
  high-coercivity, 175–80, 182
  high-permeability, initial permeability, 201
  uses, 156–9
  low-expansion, 118–19
  magnetostrictive, 160
  spontaneous magnetization, 113
  superconducting, 109–10
  with rectangular hysteresis loops, 162
Alnico alloys, 178
amber, 12, 14
ammeters, 33, 177, 185
AMPÈRE, André-Marie, 22, 29, 35
ampere, 69
AMUNDSEN, 240
ANDERSON, C. D., 226
angle of dip, 237–40, 244, 245
angular momentum, 71–3
anisotropy forces, 126
annealing, 141
antiferromagnetics, 202–4
APPLETON, Sir Edward, 247
ARAGO, D. F. J., 29
ASTON, F. W., 223
atom, 66–71
  behaviour in magnetic field, 76–8, 84–92

diamagnetic susceptibility, 91–2
electron shells, 74–5
ground and excited states, 73
interactions, 88–9
isotopes, 223–5
magnetic moment, 75, 76, 78, 83–5
paramagnetic susceptibility, 92–3
size, 70
structure, 67–70, 227
weight, 66–7, 70–1
atomic magnet, 76–7
atomic nucleus, 227
  deflection in magnetic field, 220
  magnetic moment, 217–18
atomic susceptibility, 56
  of rare gases, 93–4
atomic weights, 67

BABCOCK, H. W., 256–7
barium ferrite, 182, 200
BARKHAUSEN, H., 122
BARNETT, S. T., 100
BATES, L. F., 99, 124
batteries. *See* cells
benzine, diamagnetic susceptibility, 101–3
betatron, 232
BIOT, Jean-Baptiste, 22, 29
Bismanol, 182
bismuth, magnetic properties, 83, 105–6
BITTER, F., 123, 210

Index

BLACKETT, P. M. S., 226, 257
BLOCH, F., 131
BOHR, Niels, 69
Bohr magneton, 74
Boltzmann's constant, 86
brains, automatic, 162–5
BULLARD, Sir Edward, 253

carrier frequency, 193
cells, 19
charged particles, acceleration in electric field, 228–32
    charge-to-mass ratio, 223–5
    deflection in electric field, 222, 228
    in magnetic field, 218–26, 228
    high-energy, 227–32
CHATTOCK, A. P., 99
CHEVALLIER, R., 260–1
CLERK-MAXWELL, James, 42, 43, 44–8
closure domains, 130
cobalt, crystal structure, 81, 127–8
    single crystals, magnetic anisotropy forces, 126–8
    magnetization, 136–7, 148
cobalt ferrite, 200
coercive force, 59, 61, 144–5
    of ferrites, 200
    of permanent magnet materials, 174–83
compass, 13, 177
computers, electronic 162–6, 200
condensers, 46, 213–14
conductivity. *See* electrical conductivity
conductors, 15
    in magnetic field, force on, 32–5
    induced currents in, 41–2
    magnetic effects, 20, 27–31

continental drift, 264, 267–9
control systems, automatic and remote, 158–9
cosmic rays, 226
COULOMB, Charles, 17
crystals, diamagnetic, 95
    ferromagnetic, anistropy forces, 126
    single, magnetic properties, 124–8, 134–7
    internal stresses, 140–1
    ionic, magnetic properties, 95
    structures, 79–82
Curie temperature, 62–3, 113, 114, 117, 118, 199
Curie's Law, 57, 88–9
Curie-Weiss Law, 57, 103, 110–11
currents. *See* electric currents
cyclotrons, 228–32

DALTON, John, 66
DANIELS, J. M., 211
DEBYE, P., 215
declination, 240, 244, 247
DE HAAS, W. J., 98
demagnetization, 65
    of permanent magnets, 171–4
demagnetizing factor, 171–3
demagnetizing fields, 54, 171
demagnetizing quadrant, 172
depth sounding, 160
diamagnetic materials, 51, 58, 64, 83, 89
diamagnetic susceptibility, 83, 91–2
    of bismuth, 83, 105
    of organic compounds, 101–2
    of superconductors, 108–9
diamagnetism, 89–90
dielectric constant, 44
dipoles. *See* magnetic dipoles

Index

DIRAC, P. A. M., 76, 104, 226
domain hypothesis, 121–37
domains, behaviour in magnetic fields, 135–7, 140
  boundary movement, 135–8, 140–4, 145–6
  formation, 128–30, 134–5
  shape and size, 124, 133–5
  single, 181
  visualization, 123–4, 132
  walls, 130–3, 135
DU FAY, C. F., 15
DU TOIT, A. L., 267
dynamo-lighting for bicycles, 178
dynamos, 40–1, 152
  self-exciting, 253–5

eddy currents, 41–2
  in transformers, 153–4
  in metallic ferromagnetics, 165, 193–4
EINSTEIN, Albert, 98
electric bell, 159
electric charges, 16, 42, 68
electric currents, 18
  eddy, 41–2
    in metallic ferromagnetics, 165, 193–4
    in transformers, 153–4
  heating effects, 41, 154
  in earth's atmosphere, 247
  induced, 36–9, 41–2
  interactions, 35
  magnetic effects, 20, 27–31
  measurement, 35
  production, 40, 41
  units, 30–1, 43, 69
electric fields, 41
  associated with electromagnetic waves, 47
  force on charged particles, 222, 228–9
  magnetic effects, 43, 46
electric forces, 15, 16
  inverse square law, 16–17
electric motors, 33–6, 152
  alternating-current, 40
electric potential, 18–20
electric units, 43–4
electrical conductivity, of gases, 67–8
  of metals, 81
  of metals and alloys at low temperatures, 107–9
electricity, 71
  from chemical and mechanical energy, 39–40
  from magnetism, 36–9
electrification by friction, 14–16
electromagnet, 29, 50, 62, 205–14
electromagnetic induction, 36–9, 152
electromagnetic relays, 158–9
electromagnetic unit of pole strength, 23
electromagnetic waves, 46–7
  ultra-short, 232–3
electromagnetism, 21, 22–48
electromotive force, 37–8
  production, 40, 41, 61–2, 152, 155
electron, 68–9
  acceleration, 231, 232
  angular momentum, 72
  charge-to-mass ratio, 222
  conduction, 81, 104–5
  magnetic moment, 71
  mass, 70
  movement in electric and magnetic fields, 222, 231, 232–3
  in magnetic field, 220

Index

electron-*contd*
  orbital magnetic moment, 72–3, 74
  orbiting, effect of magnetic field, 89–91, 104–5
  spin magnetic moment, 76
  velocity, 222, 225
electron gas, 163
electron shells of atoms, 74–5
electron spin, 75–6
  in antiferromagnetics, 202–3
electrostatic unit of charge, 44
elements, chemical, 66, 75
  magnetic, 237
ELSASSER, W. M., 253
energy product of permanent magnet, 175
equator, geographical and magnetic, 244
equivalent magnetic shell, 30, 31
erg, 60
exchange energy, 114
exchange integral, 114–15, 198–9

FARADAY, Michael, 20, 33, 36–9, 50, 83, 256
FERMI, E., 104
ferrimagnetic materials in rocks, 259
ferrimagnetism, 199
ferrites, composition and physical properties, 195
  electrical properties, 196, 200
  for permanent magnets, 182, 200–1
  in computers, 164–5, 200
  magnetic properties, 196–203
  structure, 195–6
ferromagnetic particles, single-domain, 181
ferromagnetic substances, 51, 58–65, 115–16

crystal structure, 79–81
Curie temperature, 112–14, 116, 117
domain hypothesis, 121–37
eddy currents in, 193–4
effect of temperature, 62–3
in electronic computers, 162
magnetization, 99, 110–14, 121–2, 137–9
  remanent, 147–8
thermal expansion, 118–19
ferromagnetism, 104, 110–20
Ferroxcube, 201
Ferroxdure, 182, 200–1
Ferroxplana, 202
Fleming's rule, 33
flux density, 54
FRANKLIN, Benjamin, 15

GALVANI, Luigi, 19
galvanometers, 35, 185
garnets, rare earth, 200
gases, at ultra-high temperatures, 233–4
  electrical conductivity, 67–8
  magnetic properties, 93–4
  paramagnetic magnetization, 86–9
GAUSS, Karl, 54, 236
gauss, 54
GELLIBRAND, Henry, 245
GERLACH, W., 77
GIAUQUE, W. F., 215
GILBERT, William, 14, 167–8
glass-to-metal seals, 119
gramophone pick-ups, 182
GRAY, Stephen, 15
gyromagnetic effects, 99–100

HALE, G. E., 256
HEISENBERG, W., 114
HENRY, Joseph, 50
Heusler alloys, 115–16

Index

HONDA, K., 177
Hycomax, 179
hydrocarbons, aromatic, diamagnetic susceptibility, 101–2
hydrogen atom, mass, 70
  orbital magnetic moment, 74
  structure, 69–70, 72–3
hysteresis, 60, 137, 143–7
  in permanent magnets, 172
hysteresis loss, 147
  in transformers, 152–3

induction, electromagnetic, 36–9, 41–2
induction heating, 232
initial susceptibility, 64, 140–2, 145
insulators, crystal structure, 182
  electrification by friction, 14
  magnetic properties, 100–2
intensity of magnetization, 54–6
Invar, 118
inverse square law, of electric force, 16
  of magnetic force, 17, 22–3
ions, magnetic properties, 95–8, 100, 103, 104
  positive, deflection in magnetic fields, 223–5
ionosphere, 247–9, 250, 252
iron (*see also* steel)
  crystal structure, 80, 128
  magnetization, 61, 99 (*see also* ferromagnetism)
    of single crystals, 124–8, 134–6, 148
    permanent, 168
  uses, 61–2
iron–metal oxides. *See* ferrites
isogonic chart and lines, 240–1
isotopes, 223–5

JAMIN, J. C., 169–70

KAPITZA, P., 212–13
KNIGHT, Garvin, 169

LANDAU, L., 128
LANGEVIN, Paul, 83, 86
latitude, geographical and magnetic, 244
LAWRENCE, E. O., 228, 230
Le Chatelier's principle, 39
Lenz's law, 39
LIFSHITZ, E., 128
light, as electromagnetic wave, 46–7
  effect of magnetic fields, 255–6
  from excited atoms, 73
lines of force (*see also* magnetic fields)
  of bar magnet, 26
  of magnetic dipole, 27
lines of induction, 52–3
liquids, magnetic properties, 94
lodestone, 12, 15, 49, 52, 123, 167–8, 192, 195
loudspeakers, 178, 185–6

magnesium ferrite, 200
magnet (*see also* electromagnet)
  atomic, 76–7
  bar, lines of force, 26
    lines of induction, 52–3
    magnetic field, 171
    magnetic moment, 27
  demagnetizing factor, 171
  low-temperature, 210
  permanent, 58, 167–70
    care, 183–4
    design, 170–5
    early uses, 167–70
    magnetic field, 205
    magnetic induction, 175

277

Index

magnet – *contd.*
   magnetization, 183–4
   materials, 175–83, 187, 200–1
   micropowder, 182
   uses, 177–8, 182, 184–91
   spherical, magnetic field and poles, 13, 235–6
   superconducting, 109
magnetic ageing, 154
magnetic anisotrophy, 126
magnetic chuck, 187–8
magnetic compass, 13, 177
magnetic cooling, 214–17
magnetic couple, 25
magnetic deflection of charged particles, 218–26
magnetic dipoles, 27, 31–2
magnetic elements, 237
magnetic fields, 24–9
   associated with electromagnetic waves, 47
   behaviour of atoms in, 76–8, 84–92
     of domains in, 135–7, 140
     of electrons in, 89–91, 104–5
   deflection of atomic particles by, 218–26, 228, 232–3
   due to bar magnets, 26, 171
     to electric currents, 20, 27–31
     to electric fields, 43, 46
     to electromagnets, 205–14
     to magnetic shells, 30
     to permanent magnets, 205
     to spherical magnets, 13, 235–6
     to superconductors, 109
   earth's, 14, 24, 25, 235–55
     diurnal and semidiurnal variation, 246–50
     positional variation, 241–5
     rock magnetization by, 259–67
     secular variation, 245–6, 261, 262
     vector components, 237
   effects on light, 255–6
     on physical properties, 63–4
   heating effects, 233
   measurement, 258
   mechanical force on conductors in, 32–5
   moon's, 257
   production, 205–9
   pulsed, 212–14
   strength, 43
   sun's and stars', 256–7
magnetic flux, 52, 61–2
magnetic forces, 14
   inverse square law, 17, 22–3
magnetic induction, 52–5
   in permanent magnets, 175
   in superconductors, 107–8
magnetic materials, classification, 51, 204
   hard, uses, 166–70, 177–8, 182
   soft, uses, 157–68
magnetic memory arrays, 162–5, 200
magnetic moments, 27
   of atomic nuclei, 217–18
   of atoms, 75, 76, 78, 83–5
   of ferrites, 197–8
   of ions, 97–8
   orbital, 71, 72–3, 74, 98–9
   spin, 76
magnetic permeability, 24, 63
   initial, 155–7
   of various materials, 201
magnetic poles, 13–14, 22, 31–2
   earth's, 14, 240, 241, 244
     positional variation, 261, 263–4, 265–7
   of spherical magnets, 13
   strength, 23, 56

Index

magnetic screening, 158
magnetic shells, 30–1
magnetic storms, 251–2
magnetic susceptibility, 56–8 (*see also* diamagnetic susceptibility *and* paramagnetic susceptibility)
  initial, 64, 140–2, 145
  of antiferromagnetics, 202–3
  of atoms, 92
  of diamagnetics and paramagnetics, 83, 86
  of elements, 105
  of organic compounds, 101
magnetic tape-recording, 189–91
magnetism, conversion into electricity, 36–9
  induced, 49–50
  of rocks, 258–69
magnetite, 195, 197–8, 200
magnetization, by mechanical stresses, 138–9, 140–1
  by rotation, 100
  by weak fields, 136–7
  induced, 49–50, 58–61
  intensity, 54–6
  of ferromagnetics, 99, 110–14, 121–2, 137–9
  of iron and steel, 61, 99, 168
  of magnetic tape, 190
  of paramagnetics, 84–9, 98–100, 103–4, 106, 215
  of permanent magnets, 183–4
  of rocks, 259–67
  of single crystals, 124–8, 134–7
  of watches, 64–5
  saturation, 59, 61, 87, 97, 121–2, 148, 198
  spontaneous, 112–14, 117–18, 121–2, 199
  thermo-remanent, 259

magnetometers, 258
magneton, Bohr, 74
magnetostriction, 64, 138–9
  in ferrites, 200
magnetostrictive devices, 159–60
magnetron, 232–3
manganese bismuthide, 182
manganese–zinc ferrite, 201
masers, 270
mass spectrometers, 223–4
mass susceptibility, 56
MAXWELL. *See* Clerk-Maxwell
memory devices, 160–2, 200
meridian, geographical and magnetic, 241
mesons, magnetic deflection, 218, 226
metals, electrical conductivity, 81
  at low temperatures, 107–9
  induction heating, 233
  magnetic properties, 102–9
MILLIKAN, R. A., 68
MISHIMA, I., 177
MITCHELL, John, 17
molecular susceptibility, 56
molecular weights, 67
molecules, 66–7
moments. *See* magnetic moments

NÉEL, L., 195, 198–9, 263
Néel temperature, 202
neutrino, 225
nickel, coercive force, 175–6
  crystal structure, 81, 128
  initial permeability, 175–6
  magnetic anisotropic forces of single crystal, 126, 128
  magnetization, 61
  of single crystal, 136
  uses, 159–62
nuclear magnetic moments, 217–18

Index

OERSTED, Hans, 20, 28
oersted, 24
organic compounds, magnetic properties, 101–2

paramagnetic substances, 51, 57, 64
  ferromagnetic behaviour, 106
  magnetic susceptibility, 83, 92–3, 110–11
  magnetization, 84–9, 98–100, 103–4, 106, 215
  paramagnetic susceptibility, 83, 92–3, 110–11
α-particles, acceleration, 227
PASCAL, P., 101
pendulums, low-expansion, 118
PEREGRINUS DE MARICOURT, 13
periodic table of elements, 75
Permalloys, 157, 158
permeability. See magnetic permeability
Planck's constant of action, 72
POISSON, Simeon, 18
pole. See magnetic poles
positrons, 226
potential. See electric potential
POULSEN, V., 189
protons, 70
  magnetic deflection, 220
  negative, 226

radar, 232–3
radio receivers, 193, 210
radio transmission, 193
radio waves, deviation by ionosphere, 248–9
rare earths, electron configuration, 96
  magnetic properties, 97–8, 100, 103–4

rays, cosmic, 226
β-rays, magnetic deflection, 225
γ-rays, energy, 225
X-rays, ultra-short-wave, 232
record-player pick-ups, 182
rectifiers, 41
remanence, 58, 61
  in ferromagnetics, 144, 147–8
  in paramagnetics, 87
  in permanent magnets, 171, 174–5
ROWLAND, H. A., 43
RUTHERFORD, Lord, 69, 227

SAVART, Felix, 22, 29
semiconductors, 197
SHOCKLEY, W., 135
SNOEK, J. L., 194
solar flares, 251–2
solenoids, 206, 209–13
solids, magnetic properties, 94–109
sound transmission, 192–3
sound waves, high-frequency, production, 64, 159–60
  uses, 160
spectrometer, mass, 223–4
  β-ray, 225
speed of response of computers, 164–5
spinel structure, 195–6
  inverse, 197
steel (*see also* alloys *and* iron)
  for permanent magnets, 170, 176–7, 187
  for transformers, 154–5
  permanent magnetization, 168
STERN, O., 77
STEWART, Balfour, 247
STONER, E. C., 181
STURGEON, William, 50
SUCKSMITH, W., 99

Index

sunspots, 249–52
Super Invar, 118
superconductors, 107–9
Supermalloy, 157
susceptibility. *See* diamagnetic susceptibility, magnetic susceptibility *and* paramagnetic susceptibility
Swedish Mining Compass, 258
synchrotrons, 231

TAKEI, T., 177
tape-recording, magnetic, 189–91
Telegraphone, 189
telephone earpieces, 178, 187
telephone exchanges, automatic, 159, 162
television receivers, 158, 201
temperatures, very high, production, 233–4
  very low, measurement, 217
  production, 214–17, 218
THELLIER, E., 261
thermostats, 119
THOMSON, Sir J. J., 67, 68, 221–3
Ticonal, 179
TILAS, Daniel, 258

transformers, 41, 152–7, 193–4, 201
turning moment, 25, 27

ultrasonics, production and uses, 159–60
unit magnetic pole, 23
unit of electric current, 30–1
  of magnetic field strength, 24

Vectolite, 182, 200
VOLTA, Alessandro, 19
voltmeters, 177, 185
volume susceptibility, 56

watch springs, magnetization and demagnetization, 64–5
WEGENER, A., 264, 267
WEISS, Pierre, 110, 114, 121
WILLIAMS, H. J., 124, 135
Wilson cloud chamber, 226
wireless waves, 47
  high-frequency, 193
WOHLFARTH, E. P., 181

ZEEMAN, Pieter, 256
Zeeman effect, 256–7

A CATALOGUE OF SELECTED DOVER BOOKS
IN ALL FIELDS OF INTEREST

A CATALOGUE OF SELECTED DOVER BOOKS IN ALL FIELDS OF INTEREST

WHAT IS SCIENCE?, *N. Campbell*
The role of experiment and measurement, the function of mathematics, the nature of scientific laws, the difference between laws and theories, the limitations of science, and many similarly provocative topics are treated clearly and without technicalities by an eminent scientist. "Still an excellent introduction to scientific philosophy," H. Margenau in *Physics Today*. "A first-rate primer . . . deserves a wide audience," *Scientific American*. 192pp. 5⅜ x 8.
60043-2 Paperbound $1.25

THE NATURE OF LIGHT AND COLOUR IN THE OPEN AIR, *M. Minnaert*
Why are shadows sometimes blue, sometimes green, or other colors depending on the light and surroundings? What causes mirages? Why do multiple suns and moons appear in the sky? Professor Minnaert explains these unusual phenomena and hundreds of others in simple, easy-to-understand terms based on optical laws and the properties of light and color. No mathematics is required but artists, scientists, students, and everyone fascinated by these "tricks" of nature will find thousands of useful and amazing pieces of information. Hundreds of observational experiments are suggested which require no special equipment. 200 illustrations; 42 photos. xvi + 362pp. 5⅜ x 8.
20196-1 Paperbound $2.75

THE STRANGE STORY OF THE QUANTUM, AN ACCOUNT FOR THE GENERAL READER OF THE GROWTH OF IDEAS UNDERLYING OUR PRESENT ATOMIC KNOWLEDGE, *B. Hoffmann*
Presents lucidly and expertly, with barest amount of mathematics, the problems and theories which led to modern quantum physics. Dr. Hoffmann begins with the closing years of the 19th century, when certain trifling discrepancies were noticed, and with illuminating analogies and examples takes you through the brilliant concepts of Planck, Einstein, Pauli, Broglie, Bohr, Schroedinger, Heisenberg, Dirac, Sommerfeld, Feynman, etc. This edition includes a new, long postscript carrying the story through 1958. "Of the books attempting an account of the history and contents of our modern atomic physics which have come to my attention, this is the best," H. Margenau, Yale University, in *American Journal of Physics*. 32 tables and line illustrations. Index. 275pp. 5⅜ x 8.
20518-5 Paperbound $2.00

GREAT IDEAS OF MODERN MATHEMATICS: THEIR NATURE AND USE, *Jagjit Singh*
Reader with only high school math will understand main mathematical ideas of modern physics, astronomy, genetics, psychology, evolution, etc. better than many who use them as tools, but comprehend little of their basic structure. Author uses his wide knowledge of non-mathematical fields in brilliant exposition of differential equations, matrices, group theory, logic, statistics, problems of mathematical foundations, imaginary numbers, vectors, etc. Original publication. 2 appendixes. 2 indexes. 65 ills. 322pp. 5⅜ x 8.
20587-8 Paperbound $2.50

CATALOGUE OF DOVER BOOKS

THE MUSIC OF THE SPHERES: THE MATERIAL UNIVERSE — FROM ATOM TO QUASAR, SIMPLY EXPLAINED, *Guy Murchie*
Vast compendium of fact, modern concept and theory, observed and calculated data, historical background guides intelligent layman through the material universe. Brilliant exposition of earth's construction, explanations for moon's craters, atmospheric components of Venus and Mars (with data from recent fly-by's), sun spots, sequences of star birth and death, neighboring galaxies, contributions of Galileo, Tycho Brahe, Kepler, etc.; and (Vol. 2) construction of the atom (describing newly discovered sigma and xi subatomic particles), theories of sound, color and light, space and time, including relativity theory, quantum theory, wave theory, probability theory, work of Newton, Maxwell, Faraday, Einstein, de Broglie, etc. "Best presentation yet offered to the intelligent general reader," *Saturday Review*. Revised (1967). Index. 319 illustrations by the author. Total of xx + 644pp. 5⅜ x 8½.
21809-0, 21810-4 Two volume set, paperbound $5.00

FOUR LECTURES ON RELATIVITY AND SPACE, *Charles Proteus Steinmetz*
Lecture series, given by great mathematician and electrical engineer, generally considered one of the best popular-level expositions of special and general relativity theories and related questions. Steinmetz translates complex mathematical reasoning into language accessible to laymen through analogy, example and comparison. Among topics covered are relativity of motion, location, time; of mass; acceleration; 4-dimensional time-space; geometry of the gravitational field; curvature and bending of space; non-Euclidean geometry. Index. 40 illustrations. x + 142pp. 5⅜ x 8½.
61771-8 Paperbound $1.50

HOW TO KNOW THE WILD FLOWERS, *Mrs. William Starr Dana*
Classic nature book that has introduced thousands to wonders of American wild flowers. Color-season principle of organization is easy to use, even by those with no botanical training, and the genial, refreshing discussions of history, folklore, uses of over 1,000 native and escape flowers, foliage plants are informative as well as fun to read. Over 170 full-page plates, collected from several editions, may be colored in to make permanent records of finds. Revised to conform with 1950 edition of Gray's Manual of Botany. xlii + 438pp. 5⅜ x 8½.
20332-8 Paperbound $2.50

MANUAL OF THE TREES OF NORTH AMERICA, *Charles Sprague Sargent*
Still unsurpassed as most comprehensive, reliable study of North American tree characteristics, precise locations and distribution. By dean of American dendrologists. Every tree native to U.S., Canada, Alaska; 185 genera, 717 species, described in detail—leaves, flowers, fruit, winterbuds, bark, wood, growth habits, etc. plus discussion of varieties and local variants, immaturity variations. Over 100 keys, including unusual 11-page analytical key to genera, aid in identification. 783 clear illustrations of flowers, fruit, leaves. An unmatched permanent reference work for all nature lovers. Second enlarged (1926) edition. Synopsis of families. Analytical key to genera. Glossary of technical terms. Index. 783 illustrations, 1 map. Total of 982pp. 5⅜ x 8.
20277-1, 20278-X Two volume set, paperbound $6.00

CATALOGUE OF DOVER BOOKS

IT'S FUN TO MAKE THINGS FROM SCRAP MATERIALS,
Evelyn Glantz Hershoff
What use are empty spools, tin cans, bottle tops? What can be made from rubber bands, clothes pins, paper clips, and buttons? This book provides simply worded instructions and large diagrams showing you how to make cookie cutters, toy trucks, paper turkeys, Halloween masks, telephone sets, aprons, linoleum block- and spatter prints — in all 399 projects! Many are easy enough for young children to figure out for themselves; some challenging enough to entertain adults; all are remarkably ingenious ways to make things from materials that cost pennies or less! Formerly "Scrap Fun for Everyone." Index. 214 illustrations. 373pp. 5⅜ x 8½.   21251-3 Paperbound $2.00

SYMBOLIC LOGIC and THE GAME OF LOGIC, *Lewis Carroll*
"Symbolic Logic" is not concerned with modern symbolic logic, but is instead a collection of over 380 problems posed with charm and imagination, using the syllogism and a fascinating diagrammatic method of drawing conclusions. In "The Game of Logic" Carroll's whimsical imagination devises a logical game played with 2 diagrams and counters (included) to manipulate hundreds of tricky syllogisms. The final section, "Hit or Miss" is a lagniappe of 101 additional puzzles in the delightful Carroll manner. Until this reprint edition, both of these books were rarities costing up to $15 each. Symbolic Logic: Index. xxxi + 199pp. The Game of Logic: 96pp. 2 vols. bound as one. 5⅜ x 8.
20492-8 Paperbound $2.50

MATHEMATICAL PUZZLES OF SAM LOYD, PART I
selected and edited by M. Gardner
Choice puzzles by the greatest American puzzle creator and innovator. Selected from his famous collection, "Cyclopedia of Puzzles," they retain the unique style and historical flavor of the originals. There are posers based on arithmetic, algebra, probability, game theory, route tracing, topology, counter and sliding block, operations research, geometrical dissection. Includes the famous "14-15" puzzle which was a national craze, and his "Horse of a Different Color" which sold millions of copies. 117 of his most ingenious puzzles in all. 120 line drawings and diagrams. Solutions. Selected references. xx + 167pp. 5⅜ x 8.
20498-7 Paperbound $1.35

STRING FIGURES AND HOW TO MAKE THEM, *Caroline Furness Jayne*
107 string figures plus variations selected from the best primitive and modern examples developed by Navajo, Apache, pygmies of Africa, Eskimo, in Europe, Australia, China, etc. The most readily understandable, easy-to-follow book in English on perennially popular recreation. Crystal-clear exposition; step-by-step diagrams. Everyone from kindergarten children to adults looking for unusual diversion will be endlessly amused. Index. Bibliography. Introduction by A. C. Haddon. 17 full-page plates, 960 illustrations. xxiii + 401pp. 5⅜ x 8½.
20152-X Paperbound $2.50

PAPER FOLDING FOR BEGINNERS, *W. D. Murray and F. J. Rigney*
A delightful introduction to the varied and entertaining Japanese art of origami (paper folding), with a full, crystal-clear text that anticipates every difficulty; over 275 clearly labeled diagrams of all important stages in creation. You get results at each stage, since complex figures are logically developed from simpler ones. 43 different pieces are explained: sailboats, frogs, roosters, etc. 6 photographic plates. 279 diagrams. 95pp. 5⅜ x 8⅜.
20713-7 Paperbound $1.00

CATALOGUE OF DOVER BOOKS

PRINCIPLES OF ART HISTORY,
H. Wölfflin
Analyzing such terms as "baroque," "classic," "neoclassic," "primitive," "picturesque," and 164 different works by artists like Botticelli, van Cleve, Dürer, Hobbema, Holbein, Hals, Rembrandt, Titian, Brueghel, Vermeer, and many others, the author establishes the classifications of art history and style on a firm, concrete basis. This classic of art criticism shows what really occurred between the 14th-century primitives and the sophistication of the 18th century in terms of basic attitudes and philosophies. "A remarkable lesson in the art of seeing," *Sat. Rev. of Literature*. Translated from the 7th German edition. 150 illustrations. 254pp. $6\frac{1}{8}$ x $9\frac{1}{4}$. 20276-3 Paperbound $2.50

PRIMITIVE ART,
Franz Boas
This authoritative and exhaustive work by a great American anthropologist covers the entire gamut of primitive art. Pottery, leatherwork, metal work, stone work, wood, basketry, are treated in detail. Theories of primitive art, historical depth in art history, technical virtuosity, unconscious levels of patterning, symbolism, styles, literature, music, dance, etc. A must book for the interested layman, the anthropologist, artist, handicrafter (hundreds of unusual motifs), and the historian. Over 900 illustrations (50 ceramic vessels, 12 totem poles, etc.). 376pp. $5\frac{3}{8}$ x 8. 20025-6 Paperbound $2.50

THE GENTLEMAN AND CABINET MAKER'S DIRECTOR,
Thomas Chippendale
A reprint of the 1762 catalogue of furniture designs that went on to influence generations of English and Colonial and Early Republic American furniture makers. The 200 plates, most of them full-page sized, show Chippendale's designs for French (Louis XV), Gothic, and Chinese-manner chairs, sofas, canopy and dome beds, cornices, chamber organs, cabinets, shaving tables, commodes, picture frames, frets, candle stands, chimney pieces, decorations, etc. The drawings are all elegant and highly detailed; many include construction diagrams and elevations. A supplement of 24 photographs shows surviving pieces of original and Chippendale-style pieces of furniture. Brief biography of Chippendale by N. I. Bienenstock, editor of *Furniture World*. Reproduced from the 1762 edition. 200 plates, plus 19 photographic plates. vi + 249pp. $9\frac{1}{8}$ x $12\frac{1}{4}$. 21601-2 Paperbound $4.00

AMERICAN ANTIQUE FURNITURE: A BOOK FOR AMATEURS,
Edgar G. Miller, Jr.
Standard introduction and practical guide to identification of valuable American antique furniture. 2115 illustrations, mostly photographs taken by the author in 148 private homes, are arranged in chronological order in extensive chapters on chairs, sofas, chests, desks, bedsteads, mirrors, tables, clocks, and other articles. Focus is on furniture accessible to the collector, including simpler pieces and a larger than usual coverage of Empire style. Introductory chapters identify structural elements, characteristics of various styles, how to avoid fakes, etc. "We are frequently asked to name some book on American furniture that will meet the requirements of the novice collector, the beginning dealer, and . . . the general public. . . . We believe Mr. Miller's two volumes more completely satisfy this specification than any other work," *Antiques*. Appendix. Index. Total of vi + 1106pp. $7\frac{7}{8}$ x $10\frac{3}{4}$.
21599-7, 21600-4 Two volume set, paperbound $10.00

CATALOGUE OF DOVER BOOKS

THE BAD CHILD'S BOOK OF BEASTS, MORE BEASTS FOR WORSE CHILDREN, and A MORAL ALPHABET, *H. Belloc*
Hardly and anthology of humorous verse has appeared in the last 50 years without at least a couple of these famous nonsense verses. But one must see the entire volumes — with all the delightful original illustrations by Sir Basil Blackwood — to appreciate fully Belloc's charming and witty verses that play so subacidly on the platitudes of life and morals that beset his day — and ours. A great humor classic. Three books in one. Total of 157pp. 5⅜ x 8.
20749-8 Paperbound $1.25

THE DEVIL'S DICTIONARY, *Ambrose Bierce*
Sardonic and irreverent barbs puncturing the pomposities and absurdities of American politics, business, religion, literature, and arts, by the country's greatest satirist in the classic tradition. Epigrammatic as Shaw, piercing as Swift, American as Mark Twain, Will Rogers, and Fred Allen, Bierce will always remain the favorite of a small coterie of enthusiasts, and of writers and speakers whom he supplies with "some of the most gorgeous witticisms of the English language" (H. L. Mencken). Over 1000 entries in alphabetical order. 144pp. 5⅜ x 8.
20487-1 Paperbound $1.25

THE COMPLETE NONSENSE OF EDWARD LEAR.
This is the only complete edition of this master of gentle madness available at a popular price. *A Book of Nonsense, Nonsense Songs, More Nonsense Songs and Stories* in their entirety with all the old favorites that have delighted children and adults for years. The Dong With A Luminous Nose, The Jumblies, The Owl and the Pussycat, and hundreds of other bits of wonderful nonsense. 214 limericks, 3 sets of Nonsense Botany, 5 Nonsense Alphabets, 546 drawings by Lear himself, and much more. 320pp. 5⅜ x 8. 20167-8 Paperbound $1.75

THE WIT AND HUMOR OF OSCAR WILDE, *ed. by Alvin Redman*
Wilde at his most brilliant, in 1000 epigrams exposing weaknesses and hypocrisies of "civilized" society. Divided into 49 categories—sin, wealth, women, America, etc.—to aid writers, speakers. Includes excerpts from his trials, books, plays, criticism. Formerly "The Epigrams of Oscar Wilde." Introduction by Vyvyan Holland, Wilde's only living son. Introductory essay by editor. 260pp. 5⅜ x 8.
20602-5 Paperbound $1.50

A CHILD'S PRIMER OF NATURAL HISTORY, *Oliver Herford*
Scarcely an anthology of whimsy and humor has appeared in the last 50 years without a contribution from Oliver Herford. Yet the works from which these examples are drawn have been almost impossible to obtain! Here at last are Herford's improbable definitions of a menagerie of familiar and weird animals, each verse illustrated by the author's own drawings. 24 drawings in 2 colors; 24 additional drawings. vii + 95pp. 6½ x 6.
21647-0 Paperbound $1.00

THE BROWNIES: THEIR BOOK, *Palmer Cox*
The book that made the Brownies a household word. Generations of readers have enjoyed the antics, predicaments and adventures of these jovial sprites, who emerge from the forest at night to play or to come to the aid of a deserving human. Delightful illustrations by the author decorate nearly every page. 24 short verse tales with 266 illustrations. 155pp. 6⅝ x 9¼.
21265-3 Paperbound $1.50

CATALOGUE OF DOVER BOOKS

THE PRINCIPLES OF PSYCHOLOGY,
William James
The full long-course, unabridged, of one of the great classics of Western literature and science. Wonderfully lucid descriptions of human mental activity, the stream of thought, consciousness, time perception, memory, imagination, emotions, reason, abnormal phenomena, and similar topics. Original contributions are integrated with the work of such men as Berkeley, Binet, Mills, Darwin, Hume, Kant, Royce, Schopenhauer, Spinoza, Locke, Descartes, Galton, Wundt, Lotze, Herbart, Fechner, and scores of others. All contrasting interpretations of mental phenomena are examined in detail—introspective analysis, philosophical interpretation, and experimental research. "A classic," *Journal of Consulting Psychology*. "The main lines are as valid as ever," *Psychoanalytical Quarterly*. "Standard reading...a classic of interpretation," *Psychiatric Quarterly*. 94 illustrations. 1408pp. 5⅜ x 8.
20381-6, 20382-4 Two volume set, paperbound $6.00

VISUAL ILLUSIONS: THEIR CAUSES, CHARACTERISTICS AND APPLICATIONS,
M. Luckiesh
"Seeing is deceiving," asserts the author of this introduction to virtually every type of optical illusion known. The text both describes and explains the principles involved in color illusions, figure-ground, distance illusions, etc. 100 photographs, drawings and diagrams prove how easy it is to fool the sense: circles that aren't round, parallel lines that seem to bend, stationary figures that seem to move as you stare at them — illustration after illustration strains our credulity at what we see. Fascinating book from many points of view, from applications for artists, in camouflage, etc. to the psychology of vision. New introduction by William Ittleson, Dept. of Psychology, Queens College. Index. Bibliography. xxi + 252pp. 5⅜ x 8½. 21530-X Paperbound $1.75

FADS AND FALLACIES IN THE NAME OF SCIENCE,
Martin Gardner
This is the standard account of various cults, quack systems, and delusions which have masqueraded as science: hollow earth fanatics. Reich and orgone sex energy, dianetics, Atlantis, multiple moons, Forteanism, flying saucers, medical fallacies like iridiagnosis, zone therapy, etc. A new chapter has been added on Bridey Murphy, psionics, and other recent manifestations in this field. This is a fair, reasoned appraisal of eccentric theory which provides excellent inoculation against cleverly masked nonsense. "Should be read by everyone, scientist and non-scientist alike," R. T. Birge, Prof. Emeritus of Physics, Univ. of California; Former President, American Physical Society. Index. x + 365pp. 5⅜ x 8. 20394-8 Paperbound $2.00

ILLUSIONS AND DELUSIONS OF THE SUPERNATURAL AND THE OCCULT,
D. H. Rawcliffe
Holds up to rational examination hundreds of persistent delusions including crystal gazing, automatic writing, table turning, mediumistic trances, mental healing, stigmata, lycanthropy, live burial, the Indian Rope Trick, spiritualism, dowsing, telepathy, clairvoyance, ghosts, ESP, etc. The author explains and exposes the mental and physical deceptions involved, making this not only an exposé of supernatural phenomena, but a valuable exposition of characteristic types of abnormal psychology. Originally titled "The Psychology of the Occult." 14 illustrations. Index. 551pp. 5⅜ x 8. 20503-7 Paperbound $3.50

CATALOGUE OF DOVER BOOKS

FAIRY TALE COLLECTIONS, *edited by Andrew Lang*
Andrew Lang's fairy tale collections make up the richest shelf-full of traditional children's stories anywhere available. Lang supervised the translation of stories from all over the world—familiar European tales collected by Grimm, animal stories from Negro Africa, myths of primitive Australia, stories from Russia, Hungary, Iceland, Japan, and many other countries. Lang's selection of translations are unusually high; many authorities consider that the most familiar tales find their best versions in these volumes. All collections are richly decorated and illustrated by H. J. Ford and other artists.

THE BLUE FAIRY BOOK. 37 stories. 138 illustrations. ix + 390pp. 5⅜ x 8½.
21437-0 Paperbound $1.95

THE GREEN FAIRY BOOK. 42 stories. 100 illustrations. xiii + 366pp. 5⅜ x 8½.
21439-7 Paperbound $2.00

THE BROWN FAIRY BOOK. 32 stories. 50 illustrations, 8 in color. xii + 350pp. 5⅜ x 8½.
21438-9 Paperbound $1.95

THE BEST TALES OF HOFFMANN, *edited by E. F. Bleiler*
10 stories by E. T. A. Hoffmann, one of the greatest of all writers of fantasy. The tales include "The Golden Flower Pot," "Automata," "A New Year's Eve Adventure," "Nutcracker and the King of Mice," "Sand-Man," and others. Vigorous characterizations of highly eccentric personalities, remarkably imaginative situations, and intensely fast pacing has made these tales popular all over the world for 150 years. Editor's introduction. 7 drawings by Hoffmann. xxxiii + 419pp. 5⅜ x 8½.
21793-0 Paperbound $2.25

GHOST AND HORROR STORIES OF AMBROSE BIERCE,
edited by E. F. Bleiler
Morbid, eerie, horrifying tales of possessed poets, shabby aristocrats, revived corpses, and haunted malefactors. Widely acknowledged as the best of their kind between Poe and the moderns, reflecting their author's inner torment and bitter view of life. Includes "Damned Thing," "The Middle Toe of the Right Foot," "The Eyes of the Panther," "Visions of the Night," "Moxon's Master," and over a dozen others. Editor's introduction. xxii + 199pp. 5⅜ x 8½.
20767-6 Paperbound $1.50

THREE GOTHIC NOVELS, *edited by E. F. Bleiler*
Originators of the still popular Gothic novel form, influential in ushering in early 19th-century Romanticism. Horace Walpole's *Castle of Otranto*, William Beckford's *Vathek*, John Polidori's *The Vampyre*, and a *Fragment* by Lord Byron are enjoyable as exciting reading or as documents in the history of English literature. Editor's introduction. xi + 291pp. 5⅜ x 8½.
21232-7 Paperbound $2.00

BEST GHOST STORIES OF LEFANU, *edited by E. F. Bleiler*
Though admired by such critics as V. S. Pritchett, Charles Dickens and Henry James, ghost stories by the Irish novelist Joseph Sheridan LeFanu have never become as widely known as his detective fiction. About half of the 16 stories in this collection have never before been available in America. Collection includes "Carmilla" (perhaps the best vampire story ever written), "The Haunted Baronet," "The Fortunes of Sir Robert Ardagh," and the classic "Green Tea." Editor's introduction. 7 contemporary illustrations. Portrait of LeFanu. xii + 467pp. 5⅜ x 8.
20415-4 Paperbound $2.50

CATALOGUE OF DOVER BOOKS

EASY-TO-DO ENTERTAINMENTS AND DIVERSIONS WITH COINS, CARDS, STRING, PAPER AND MATCHES, *R. M. Abraham*
Over 300 tricks, games and puzzles will provide young readers with absorbing fun. Sections on card games; paper-folding; tricks with coins, matches and pieces of string; games for the agile; toy-making from common household objects; mathematical recreations; and 50 miscellaneous pastimes. Anyone in charge of groups of youngsters, including hard-pressed parents, and in need of suggestions on how to keep children sensibly amused and quietly content will find this book indispensable. Clear, simple text, copious number of delightful line drawings and illustrative diagrams. Originally titled "Winter Nights' Entertainments." Introduction by Lord Baden Powell. 329 illustrations. v + 186pp. 5⅜ x 8½. 20921-0 Paperbound $1.25

AN INTRODUCTION TO CHESS MOVES AND TACTICS SIMPLY EXPLAINED, *Leonard Barden*
Beginner's introduction to the royal game. Names, possible moves of the pieces, definitions of essential terms, how games are won, etc. explained in 30-odd pages. With this background you'll be able to sit right down and play. Balance of book teaches strategy — openings, middle game, typical endgame play, and suggestions for improving your game. A sample game is fully analyzed. True middle-level introduction, teaching you all the essentials without oversimplifying or losing you in a maze of detail. 58 figures. 102pp. 5⅜ x 8½. 21210-6 Paperbound $1.25

LASKER'S MANUAL OF CHESS, *Dr. Emanuel Lasker*
Probably the greatest chess player of modern times, Dr. Emanuel Lasker held the world championship 28 years, independent of passing schools or fashions. This unmatched study of the game, chiefly for intermediate to skilled players, analyzes basic methods, combinations, position play, the aesthetics of chess, dozens of different openings, etc., with constant reference to great modern games. Contains a brilliant exposition of Steinitz's important theories. Introduction by Fred Reinfeld. Tables of Lasker's tournament record. 3 indices. 308 diagrams. 1 photograph. xxx + 349pp. 5⅜ x 8. 20640-8 Paperbound $2.50

COMBINATIONS: THE HEART OF CHESS, *Irving Chernev*
Step-by-step from simple combinations to complex, this book, by a well-known chess writer, shows you the intricacies of pins, counter-pins, knight forks, and smothered mates. Other chapters show alternate lines of play to those taken in actual championship games; boomerang combinations; classic examples of brilliant combination play by Nimzovich, Rubinstein, Tarrasch, Botvinnik, Alekhine and Capablanca. Index. 356 diagrams. ix + 245pp. 5⅜ x 8½. 21744-2 Paperbound $2.00

HOW TO SOLVE CHESS PROBLEMS, *K. S. Howard*
Full of practical suggestions for the fan or the beginner — who knows only the moves of the chessmen. Contains preliminary section and 58 two-move, 46 three-move, and 8 four-move problems composed by 27 outstanding American problem creators in the last 30 years. Explanation of all terms and exhaustive index. "Just what is wanted for the student," Brian Harley. 112 problems, solutions. vi + 171pp. 5⅜ x 8. 20748-X Paperbound $1.50

SOCIAL THOUGHT FROM LORE TO SCIENCE, H. E. Barnes and H. Becker

An immense survey of sociological thought and ways of viewing, studying, planning, and reforming society from earliest times to the present. Includes thought on society of preliterate peoples, ancient non-Western cultures, and every great movement in Europe, America, and modern Japan. Analyzes hundreds of great thinkers: Plato, Augustine, Bodin, Vico, Montesquieu, Herder, Comte, Marx, etc. Weighs the contributions of utopians, sophists, fascists and communists; economists, jurists, philosophers, ecclesiastics, and every 19th and 20th century school of scientific sociology, anthropology, and social psychology throughout the world. Combines topical, chronological, and regional approaches, treating the evolution of social thought as a process rather than as a series of mere topics. "Impressive accuracy, competence, and discrimination . . . easily the best single survey," Nation. Thoroughly revised, with new material up to 1960. 2 indexes. Over 2200 bibliographical notes. Three volume set. Total of 1586pp. 5⅜ x 8.
20901-6, 20902-4, 20903-2 Three volume set, paperbound $10.50

A HISTORY OF HISTORICAL WRITING, Harry Elmer Barnes

Virtually the only adequate survey of the whole course of historical writing in a single volume. Surveys developments from the beginnings of historiography in the ancient Near East and the Classical World, up through the Cold War. Covers major historians in detail, shows interrelationship with cultural background, makes clear individual contributions, evaluates and estimates importance; also enormously rich upon minor authors and thinkers who are usually passed over. Packed with scholarship and learning, clear, easily written. Indispensable to every student of history. Revised and enlarged up to 1961. Index and bibliography. xv + 442pp. 5⅜ x 8½.
20104-X Paperbound $3.00

JOHANN SEBASTIAN BACH, Philipp Spitta

The complete and unabridged text of the definitive study of Bach. Written some 70 years ago, it is still unsurpassed for its coverage of nearly all aspects of Bach's life and work. There could hardly be a finer non-technical introduction to Bach's music than the detailed, lucid analyses which Spitta provides for hundreds of individual pieces. 26 solid pages are devoted to the B minor mass, for example, and 30 pages to the glorious St. Matthew Passion. This monumental set also includes a major analysis of the music of the 18th century: Buxtehude, Pachelbel, etc. "Unchallenged as the last word on one of the supreme geniuses of music," John Barkham, Saturday Review Syndicate. Total of 1819pp. Heavy cloth binding. 5⅜ x 8.
22278-0, 22279-9 Two volume set, clothbound $15.00

BEETHOVEN AND HIS NINE SYMPHONIES, George Grove

In this modern middle-level classic of musicology Grove not only analyzes all nine of Beethoven's symphonies very thoroughly in terms of their musical structure, but also discusses the circumstances under which they were written, Beethoven's stylistic development, and much other background material. This is an extremely rich book, yet very easily followed; it is highly recommended to anyone seriously interested in music. Over 250 musical passages. Index. viii + 407pp. 5⅜ x 8.
20334-4 Paperbound $2.50

THE TIME STREAM
John Taine

Acknowledged by many as the best SF writer of the 1920's, Taine (under the name Eric Temple Bell) was also a Professor of Mathematics of considerable renown. Reprinted here are *The Time Stream*, generally considered Taine's best, *The Greatest Game*, a biological-fiction novel, and *The Purple Sapphire*, involving a supercivilization of the past. Taine's stories tie fantastic narratives to frameworks of original and logical scientific concepts. Speculation is often profound on such questions as the nature of time, concept of entropy, cyclical universes, etc. 4 contemporary illustrations. v + 532pp. 5⅜ x 8⅜.

21180-0 Paperbound $3.00

SEVEN SCIENCE FICTION NOVELS,
H. G. Wells

Full unabridged texts of 7 science-fiction novels of the master. Ranging from biology, physics, chemistry, astronomy, to sociology and other studies, Mr. Wells extrapolates whole worlds of strange and intriguing character. "One will have to go far to match this for entertainment, excitement, and sheer pleasure . . ."*New York Times*. Contents: The Time Machine, The Island of Dr. Moreau, The First Men in the Moon, The Invisible Man, The War of the Worlds, The Food of the Gods, In The Days of the Comet. 1015pp. 5⅜ x 8.

20264-X Clothbound $5.00

28 SCIENCE FICTION STORIES OF H. G. WELLS.

Two full, unabridged novels, *Men Like Gods* and *Star Begotten*, plus 26 short stories by the master science-fiction writer of all time! Stories of space, time, invention, exploration, futuristic adventure. Partial contents: *The Country of the Blind, In the Abyss, The Crystal Egg, The Man Who Could Work Miracles, A Story of Days to Come, The Empire of the Ants, The Magic Shop, The Valley of the Spiders, A Story of the Stone Age, Under the Knife, Sea Raiders*, etc. An indispensable collection for the library of anyone interested in science fiction adventure. 928pp. 5⅜ x 8.

20265-8 Clothbound $5.00

THREE MARTIAN NOVELS,
Edgar Rice Burroughs

Complete, unabridged reprinting, in one volume, of Thuvia, Maid of Mars; Chessmen of Mars; The Master Mind of Mars. Hours of science-fiction adventure by a modern master storyteller. Reset in large clear type for easy reading. 16 illustrations by J. Allen St. John. vi + 490pp. 5⅜ x 8½.

20039-6. Paperbound $2.50

AN INTELLECTUAL AND CULTURAL HISTORY OF THE WESTERN WORLD,
Harry Elmer Barnes

Monumental 3-volume survey of intellectual development of Europe from primitive cultures to the present day. Every significant product of human intellect traced through history: art, literature, mathematics, physical sciences, medicine, music, technology, social sciences, religions, jurisprudence, education, etc. Presentation is lucid and specific, analyzing in detail specific discoveries, theories, literary works, and so on. Revised (1965) by recognized scholars in specialized fields under the direction of Prof. Barnes. Revised bibliography. Indexes. 24 illustrations. Total of xxix + 1318pp.

21275-0, 21276-9, 21277-7 Three volume set, paperbound $7.75

CATALOGUE OF DOVER BOOKS

HEAR ME TALKIN' TO YA, *edited by Nat Shapiro and Nat Hentoff*
In their own words, Louis Armstrong, King Oliver, Fletcher Henderson, Bunk Johnson, Bix Beiderbecke, Billy Holiday, Fats Waller, Jelly Roll Morton, Duke Ellington, and many others comment on the origins of jazz in New Orleans and its growth in Chicago's South Side, Kansas City's jam sessions, Depression Harlem, and the modernism of the West Coast schools. Taken from taped conversations, letters, magazine articles, other first-hand sources. Editors' introduction. xvi + 429pp. 5⅜ x 8½. 21726-4 Paperbound $2.50

THE JOURNAL OF HENRY D. THOREAU
A 25-year record by the great American observer and critic, as complete a record of a great man's inner life as is anywhere available. Thoreau's Journals served him as raw material for his formal pieces, as a place where he could develop his ideas, as an outlet for his interests in wild life and plants, in writing as an art, in classics of literature, Walt Whitman and other contemporaries, in politics, slavery, individual's relation to the State, etc. The Journals present a portrait of a remarkable man, and are an observant social history. Unabridged republication of 1906 edition, Bradford Torrey and Francis H. Allen, editors. Illustrations. Total of 1888pp. 8⅜ x 12¼.
20312-3, 20313-1 Two volume set, clothbound $30.00

A SHAKESPEARIAN GRAMMAR, *E. A. Abbott*
Basic reference to Shakespeare and his contemporaries, explaining through thousands of quotations from Shakespeare, Jonson, Beaumont and Fletcher, North's *Plutarch* and other sources the grammatical usage differing from the modern. First published in 1870 and written by a scholar who spent much of his life isolating principles of Elizabethan language, the book is unlikely ever to be superseded. Indexes. xxiv + 511pp. 5⅜ x 8½. 21582-2 Paperbound $3.00

FOLK-LORE OF SHAKESPEARE, *T. F. Thistelton Dyer*
Classic study, drawing from Shakespeare a large body of references to supernatural beliefs, terminology of falconry and hunting, games and sports, good luck charms, marriage customs, folk medicines, superstitions about plants, animals, birds, argot of the underworld, sexual slang of London, proverbs, drinking customs, weather lore, and much else. From full compilation comes a mirror of the 17th-century popular mind. Index. ix + 526pp. 5⅜ x 8½.
21614-4 Paperbound $3.25

THE NEW VARIORUM SHAKESPEARE, *edited by H. H. Furness*
By far the richest editions of the plays ever produced in any country or language. Each volume contains complete text (usually First Folio) of the play, all variants in Quarto and other Folio texts, editorial changes by every major editor to Furness's own time (1900), footnotes to obscure references or language, extensive quotes from literature of Shakespearian criticism, essays on plot sources (often reprinting sources in full), and much more.

HAMLET, *edited by H. H. Furness*
Total of xxvi + 905pp. 5⅜ x 8½.
21004-9, 21005-7 Two volume set, paperbound $5.50

TWELFTH NIGHT, *edited by H. H. Furness*
Index. xxii + 434pp. 5⅜ x 8½. 21189-4 Paperbound $2.75

LA BOHEME BY GIACOMO PUCCINI,
translated and introduced by Ellen H. Bleiler
Complete handbook for the operagoer, with everything needed for full enjoyment except the musical score itself. Complete Italian libretto, with new, modern English line-by-line translation—the only libretto printing all repeats; biography of Puccini; the librettists; background to the opera, Murger's La Boheme, etc.; circumstances of composition and performances; plot summary; and pictorial section of 73 illustrations showing Puccini, famous singers and performances, etc. Large clear type for easy reading. 124pp. 5⅜ x 8½.
20404-9 Paperbound $1.50

ANTONIO STRADIVARI: HIS LIFE AND WORK (1644-1737),
W. Henry Hill, Arthur F. Hill, and Alfred E. Hill
Still the only book that really delves into life and art of the incomparable Italian craftsman, maker of the finest musical instruments in the world today. The authors, expert violin-makers themselves, discuss Stradivari's ancestry, his construction and finishing techniques, distinguished characteristics of many of his instruments and their locations. Included, too, is story of introduction of his instruments into France, England, first revelation of their supreme merit, and information on his labels, number of instruments made, prices, mystery of ingredients of his varnish, tone of pre-1684 Stradivari violin and changes between 1684 and 1690. An extremely interesting, informative account for all music lovers, from craftsman to concert-goer. Republication of original (1902) edition. New introduction by Sydney Beck, Head of Rare Book and Manuscript Collections, Music Division, New York Public Library. Analytical index by Rembert Wurlitzer. Appendixes. 68 illustrations. 30 full-page plates. 4 in color. xxvi + 315pp. 5⅜ x 8½. 20425-1 Paperbound $3.00

MUSICAL AUTOGRAPHS FROM MONTEVERDI TO HINDEMITH,
Emanuel Winternitz
For beauty, for intrinsic interest, for perspective on the composer's personality, for subtleties of phrasing, shading, emphasis indicated in the autograph but suppressed in the printed score, the mss. of musical composition are fascinating documents which repay close study in many different ways. This 2-volume work reprints facsimiles of mss. by virtually every major composer, and many minor figures—196 examples in all. A full text points out what can be learned from mss., analyzes each sample. Index. Bibliography. 18 figures. 196 plates. Total of 170pp. of text. 7⅞ x 10¾.
21312-9, 21313-7 Two volume set, paperbound $5.00

J. S. BACH,
Albert Schweitzer
One of the few great full-length studies of Bach's life and work, and the study upon which Schweitzer's renown as a musicologist rests. On first appearance (1911), revolutionized Bach performance. The only writer on Bach to be musicologist, performing musician, and student of history, theology and philosophy, Schweitzer contributes particularly full sections on history of German Protestant church music, theories on motivic pictorial representations in vocal music, and practical suggestions for performance. Translated by Ernest Newman. Indexes. 5 illustrations. 650 musical examples. Total of xix + 928pp. 5⅜ x 8½. 21631-4, 21632-2 Two volume set, paperbound $5.00

CATALOGUE OF DOVER BOOKS

THE METHODS OF ETHICS, *Henry Sidgwick*
Propounding no organized system of its own, study subjects every major methodological approach to ethics to rigorous, objective analysis. Study discusses and relates ethical thought of Plato, Aristotle, Bentham, Clarke, Butler, Hobbes, Hume, Mill, Spencer, Kant, and dozens of others. Sidgwick retains conclusions from each system which follow from ethical premises, rejecting the faulty. Considered by many in the field to be among the most important treatises on ethical philosophy. Appendix. Index. xlvii + 528pp. 5⅜ x 8½.
21608-X Paperbound $3.00

TEUTONIC MYTHOLOGY, *Jakob Grimm*
A milestone in Western culture; the work which established on a modern basis the study of history of religions and comparative religions. 4-volume work assembles and interprets everything available on religious and folkloristic beliefs of Germanic people (including Scandinavians, Anglo-Saxons, etc.). Assembling material from such sources as Tacitus, surviving Old Norse and Icelandic texts, archeological remains, folktales, surviving superstitions, comparative traditions, linguistic analysis, etc. Grimm explores pagan deities, heroes, folklore of nature, religious practices, and every other area of pagan German belief. To this day, the unrivaled, definitive, exhaustive study. Translated by J. S. Stallybrass from 4th (1883) German edition. Indexes. Total of lxxvii + 1887pp. 5⅜ x 8½.
21602-0, 21603-9, 21604-7, 21605-5 Four volume set, paperbound $12.00

THE I CHING, *translated by James Legge*
Called "The Book of Changes" in English, this is one of the Five Classics edited by Confucius, basic and central to Chinese thought. Explains perhaps the most complex system of divination known, founded on the theory that all things happening at any one time have characteristic features which can be isolated and related. Significant in Oriental studies, in history of religions and philosophy, and also to Jungian psychoanalysis and other areas of modern European thought. Index. Appendixes. 6 plates. xxi + 448pp. 5⅜ x 8½.
21062-6 Paperbound $2.75

HISTORY OF ANCIENT PHILOSOPHY, *W. Windelband*
One of the clearest, most accurate comprehensive surveys of Greek and Roman philosophy. Discusses ancient philosophy in general, intellectual life in Greece in the 7th and 6th centuries B.C., Thales, Anaximander, Anaximenes, Heraclitus, the Eleatics, Empedocles, Anaxagoras, Leucippus, the Pythagoreans, the Sophists, Socrates, Democritus (20 pages), Plato (50 pages), Aristotle (70 pages), the Peripatetics, Stoics, Epicureans, Sceptics, Neo-platonists, Christian Apologists, etc. 2nd German edition translated by H. E. Cushman. xv + 393pp. 5⅜ x 8.
20357-3 Paperbound $3.00

THE PALACE OF PLEASURE, *William Painter*
Elizabethan versions of Italian and French novels from *The Decameron*, Cinthio, Straparola, Queen Margaret of Navarre, and other continental sources — the very work that provided Shakespeare and dozens of his contemporaries with many of their plots and sub-plots and, therefore, justly considered one of the most influential books in all English literature. It is also a book that any reader will still enjoy. Total of cviii + 1,224pp.
21691-8, 21692-6, 21693-4 Three volume set, paperbound $8.25

CATALOGUE OF DOVER BOOKS

THE WONDERFUL WIZARD OF OZ, *L. F. Baum*
All the original W. W. Denslow illustrations in full color—as much a part of "The Wizard" as Tenniel's drawings are of "Alice in Wonderland." "The Wizard" is still America's best-loved fairy tale, in which, as the author expresses it, "The wonderment and joy are retained and the heartaches and nightmares left out." Now today's young readers can enjoy every word and wonderful picture of the original book. New introduction by Martin Gardner. A Baum bibliography. 23 full-page color plates. viii + 268pp. 5⅜ x 8.
20691-2 Paperbound $1.95

THE MARVELOUS LAND OF OZ, *L. F. Baum*
This is the equally enchanting sequel to the "Wizard," continuing the adventures of the Scarecrow and the Tin Woodman. The hero this time is a little boy named Tip, and all the delightful Oz magic is still present. This is the Oz book with the Animated Saw-Horse, the Woggle-Bug, and Jack Pumpkinhead. All the original John R. Neill illustrations, 10 in full color. 287pp. 5⅜ x 8.
20692-0 Paperbound $1.75

ALICE'S ADVENTURES UNDER GROUND, *Lewis Carroll*
The original *Alice in Wonderland*, hand-lettered and illustrated by Carroll himself, and originally presented as a Christmas gift to a child-friend. Adults as well as children will enjoy this charming volume, reproduced faithfully in this Dover edition. While the story is essentially the same, there are slight changes, and Carroll's spritely drawings present an intriguing alternative to the famous Tenniel illustrations. One of the most popular books in Dover's catalogue. Introduction by Martin Gardner. 38 illustrations. 128pp. 5⅜ x 8½.
21482-6 Paperbound $1.00

THE NURSERY "ALICE," *Lewis Carroll*
While most of us consider *Alice in Wonderland* a story for children of all ages, Carroll himself felt it was beyond younger children. He therefore provided this simplified version, illustrated with the famous Tenniel drawings enlarged and colored in delicate tints, for children aged "from Nought to Five." Dover's edition of this now rare classic is a faithful copy of the 1889 printing, including 20 illustrations by Tenniel, and front and back covers reproduced in full color. Introduction by Martin Gardner. xxiii + 67pp. 6⅛ x 9¼.
21610-1 Paperbound $1.75

THE STORY OF KING ARTHUR AND HIS KNIGHTS, *Howard Pyle*
A fast-paced, exciting retelling of the best known Arthurian legends for young readers by one of America's best story tellers and illustrators. The sword Excalibur, wooing of Guinevere, Merlin and his downfall, adventures of Sir Pellias and Gawaine, and others. The pen and ink illustrations are vividly imagined and wonderfully drawn. 41 illustrations. xviii + 313pp. 6⅛ x 9¼.
21445-1 Paperbound $2.00

Prices subject to change without notice.

Available at your book dealer or write for free catalogue to Dept. Adsci, Dover Publications, Inc., 180 Varick St., N.Y., N.Y. 10014. Dover publishes more than 150 books each year on science, elementary and advanced mathematics, biology, music, art, literary history, social sciences and other areas.